5th International Conference on Biomolecular Engineering (ICBE 2015)

Lost Pines, Texas, USA
11-14 January 2015

ISBN: 978-1-5108-1576-6

Printed from e-media with permission by:

Curran Associates, Inc.
57 Morehouse Lane
Red Hook, NY 12571

Some format issues inherent in the e-media version may also appear in this print version.

Printed by Curran Associates, Inc. (2015)

For permission requests, please contact AIChE
at the address below.

AIChE
120 Wall Street, FL 23
New York, NY 10005-4020

Phone: (800) 242-4363
Fax: (203) 775-5177

www.aiche.org

Additional copies of this publication are available from:

Curran Associates, Inc.
57 Morehouse Lane
Red Hook, NY 12571 USA
Phone: 845-758-0400
Fax: 845-758-2634
Email: curran@proceedings.com
Web: www.proceedings.com

TABLE OF CONTENTS

KEYNOTE

Multipurpose Bacterial Cell Factories: New Production Strains and Novel Tools 1
Michael Bott
Non-Canonical Amino Acids as Probes of Protein Synthesis in Complex Biological Systems 2
David Tirrell
Signals from the Surface to Promote Human Pluripotent Stem Cell Differentiation 3
Laura Kiessling
Development of Microbial Cell Factories for Biorefineries 4
Akihiko Kondo

SYNTHETIC BIOLOGY

Reinforcing Synthetic Biology Against Evolutionary Failure 5
Jeffrey E. Barrick
Engineering Smarter and Stronger T Cells for Cancer Immunotherapy 6
Yvonne Chen
Synthetic Regulators for Evolutionary Metabolic Engineering 7
Gyoo Yeol Jung
Synthetic Biology with a Cell-Free TX-TL System: Metabolism, Gene Circuits and Minimal Cell in a Test Tube 8
Jonghyeon Shin, Filippo Caschera and Vincent Noireaux

METABOLIC ENGINEERING FOR FUELS AND CHEMICALS

Engineering alcohol tolerance in yeast fermentations 9
Gregory Stephanopoulos, Gerald Fink and Felix H. Lam
The Quest for Synthetic Methylotrophy 10
Eleftherios T. Papoutsakis, W. Brian Whitaker, Nicholas R. Sandoval, R. Kyle Bennett and Alan G. Fast
Metabolic Engineering of Saccharomyces Cerevisiae for the Synthesis of Polyketides and Fatty Acids 11
Nancy A. Da Silva, Christopher Leber, Javier Cardenas, Jin Wook Choi and Ruben Fernandez-Moya
Exploring P450 Expression in Escherichia coli for the Synthesis of Complex Molecules 12
Bradley Biggs, Chin Giaw Lim, Aditi Das, Marjan De Mey and Ajikumar Parayil

PROTEIN ENGINEERING AND EVOLUTION

Co-Evolution of Affinity and Stability for Domain Antibodies That Recognize Hydrophobic Antigens 13
Peter M. Tessier
Engineering Proteases to Detect Post-Translational Modifications 14
Navin Varadarajan

EXPERIMENTAL AND COMPUTATIONAL TOOLS FOR ENGINEERING BIOMOLECULES

Computational Strain Design for Improved Productivity, Yield and Robustness 16
Radhakrishnan Mahadevan
Machine-Based Learning for Rational Engineering of Biomolecular Libraries 17
Lydia M. Contreras, Jorge Vasquez, Kris Reyes and Warren Powell
De Novo Biosynthesis of Terminal Alkyne-Tagged Natural Products and Applications 18
Wenjun Zhang

Development and Analysis of Precursor Production Strains for Chemical Production 19
Xiaolin Zhang, Christopher J. Tervo and Jennifer Reed

BIOPHYSICAL MODELS OF BIOLOGICAL PROCESSES

Micromachines and motility: Single and collective cancer cell invasion 21
Muhammad H. Zaman
Multiscale Models of Antibiotic Cellbots 22
Kathryn Geldart, Brittany Forkus, Panagiota Kyriakou, Yiannis N. Kaznessis and Michail Vlysidis
Nanoscale Architecture of Tension Generation within Focal Adhesions 24
Alexander Dunn
Automated Physics-Based Design of Synthetic Riboswitches from Diverse Aptamers 25
Howard M. Salis

HIGH THROUGHPUT BIOLOGICAL DESIGN

Zymergen 26
Zach Serber
Engineering Microbial Fatty Acid Biosynthesis as an Industrial Biotechnology Platform Steve Del Cardayre* 27
Steve del Cardayre
Orthogonal Genetics for Rapid Evolution and Synthetic Biology 28
Chang C. Liu
Reverse Engineering Enzymes with High-Throughput Sequence-Function Mapping 60
Philip A. Romero and Adam R. Abate

BIOMOLECULAR DESIGN OF DIAGNOSTIC AND THERAPEUTIC AGENTS

Engineering Picomolar Affinity into a Rationally Identified 5 Kda Scaffold for Tumor Targeting 71
Max A Kruziki, Hong Zhou, Patrick Holec and Benjamin J. Hackel
Ultra-High Affinity Engineered Protein Therapeutics for Treating Metastatic Disease 73
Jennifer R. Cochran
Modular Receptor Engineering for Programming Cell-Based Therapies to Interface with Host Physiology 74
Nichole Daringer, Rachel M. Dudek, Kelly A. Schwarz and Joshua N. Leonard

DESIGNING NON-NATIVE BIOLOGICAL FUNTION

Non-Native Small-Molecule Biosensors for Screening and Selection of High-Performance Cell Factories 75
Michael K. Jensen, Mette Skjødt, Jie Zhang, Arun S. Rajkumar and Jay D. Keasling
Designing Microbes for Executive Function 76
William E. Bentley, Gregory F. Payne, Jessica Terrell, Hsuan-Chen Wu, Chen-Yu Tsao, Tanya Gordonov, Amin Zargar and David Quan
Engineering Orthogonal Translation Systems 77
Michael C. Jewett
Microbial Production of Curcuminoids 78
Nobutaka Funa

POSTER SESSION

Bias and Error Correction in Antibody Repertoire Sequencing for More Accurate Vaccine Profiling 80
Tarik Khan and Sai T. Reddy
Studies on the Influence of Different Metabolic Uncouplers on the Biodegradation of Toluene in a Differential Biofilter Reactor 81
Swaminathan Detchanamurthy and Peter Gostomski

A Biobased Approach to Mechanically Tunable Polyester 82
Kechun Zhang
Tailor Designed Peptides Bio-Inspired from Transmembrane Proteins for the Cosmetic Industry 83
Sonia Milena Aguilera Segura, Vanessa Lucia Nuñez Velez, Oscar A. Alvarez Solano, Luke Achenie, Rodrigo Torres Saez and Andres Fernando González Barrios
Energizing Yeast Cell-Free Protein Synthesis with Glucose Metabolism 84
Jessica C. Stark, Mark Anderson, C. Eric Hodgman and Michael C. Jewett
Investigation of Metabolic Capabilities of Recombinant Lactococcus Lactis for Production of Hyaluronic Acid Using Constraint Based Genome Scale Models 85
Abinaya Badri, Karthik Raman and Guhan Jayaraman
Development and in Vivo Testing of Antibodies Targeting Cancer-Associated Fibroblasts 88
James A. Van Deventer, Saravanan Rajan, Sachdev Sidhu and K. Dane Wittrup
The Bio Phantom Tissue Model 89
Jose Calderon and Carlos Calderon
In silico Prediction of Gene Deletion Targets in Escherichia coli for Enhanced Succinic Acid Production Using a Model-Guided Approach Under the Optflux Software Platform 90
Bashir S. Mienda and Mohd S. Shamsir
Bypassing Local Maxima in Protein Directed Evolution through Negative Selection 91
Barrett Steinberg and Marc Ostermeier
Controlling Cells through RNA Folding 92
Julius B. Lucks
A Human Therapeutic Enzyme Specifically Sabotages Tumor Metabolism By an Engineered Cystine/Cysteine Degrading Activity 93
Shira Cramer, Achinto Saha, Surendar Tadi, Stefano Tiziani, John Digiovanni, Everett M. Stone and George Georgiou
Glycoengineered Outer Membrane Vesicles Displaying O-Polysaccharide Antigens Elicit Protection Against Francisella Tularensis 94
Linxiao Chen, Jenny Baker, Jack Chung-Jr Huang, Christine Endicott, David Putnam, Bradley Jones and Matthew P. DeLisa
Engineering and Preclinical Development of a Human Enzyme for Cancer Therapy Via the Depletion of the Serum L-Cysteine Pool 95
Everett Stone, Shira Cramer, Achinto Saha, Stefano Tiziani, Surendar Tadi, John Digiovanni and George Georgiou
Towards Biofuel Production in Synechocystis Sp. PCC 6803: Expanding the Molecular Biology Toolbox for Pathway Engineering 96
Christie A.M. Peebles
Deciphering Transcriptional Regulation Patterns for Novel Enzyme Discovery 97
Kevin V. Solomon, Charles Haitjema, John K. Henske, Diego Borges-Rivera, Dawn A. Thompson and Michelle A. O'Malley
Design and Engineer L-Amino Acid Deaminase from Proteus vulgaris As a Robust Biocatalyst for One-Step Biosynthesis of a-Keto Acids 98
Long Liu
Reducing Uncertainty in Metabolic Networks through Thermodynamics Based Flux Analysis 99
Alexandros Kiparissides and Vassily Hatzimanikatis
Controlling Local Substrate Concentrations and Enzyme Kinetics through Rationally Designed Intermolecular Interactions 100
Yingning Gao, Jie Zhu, Jyun-Liang Lin and Ian Wheeldon
Discovery of a Pre-Eclampsia Associated Antibody That Binds to a Viral Antigen and Human Protein 101
Serra E. Elliott, Alex R. Soffici and Patrick S. Daugherty
Study of the Effects of Alternative Fluxes on the Metabolic Control Analysis of Optimally Grown E. coli 103
Geogios Fengos, Meric Ataman, Daniel Hernandez Gardiol, Ljubisa Miskovic and Vassily Hatzimanikatis
Intracellular Trafficking and Degradation of Antibody Drug Conjugates 104
Katie Maass and K. Dane Wittrup
Ligand Responsive Hybrid Promoters and Bidirectional Promoters for Metabolic Engineering in Oleaginous Yeast Yarrowia Lipolytica 105
Murtaza Shabbir-Hussain, Lauren Gambill, Samuel Williams and Mark A. Blenner
Development of SNAP-Tagged Antibodies for Magnetic Bead-Based Immuno-PCR 106
Jimmy Gollihar, Arti Pothukuchy, Michelle Byrom and Andrew Ellington
Pigment-Based, Low-Cost, Portable Micronutrient Status Tests Using Engineered Bacteria 107
Daniel Watstein and Mark P. Styczynski

High-Throughput Design of Regulatory Protein-Based Biosensors for Screening Biosynthesis Libraries 108
Patrick C. Cirino, Christopher S. Frei and Shuai Qian

Intracellular FRET-Based Assay for Redesigning the Specificity of Secreted Proteases 109
Jennifer L. Guerrero, Michelle A. O'Malley and Patrick S. Daugherty

Talens Assisted Multiplex Engineering for Accelerated Genome Evolution in Saccharomyces Cerevisiae 110
Guoqiang Zhang and Qinhong Wang

A Novel, in Vivo Continuous Evolution Strategy in Saccharomyces Cerevisiae to Create Pathway and Protein Variants 111
Jie Sun, Nathan Crook, Joe Abatemarco and Hal Alper

Generation of Protease-Inhibiting Monoclonal Antibodies By Novel Paratope Design and Function-Based Screening 112
Dong Hyun Nam, Kuili Fang, Carlos Rodriguez, Tyler Lopez and Xin Ge

Directed Protein Evolution for Specificity and Affinity Towards Chemically Modified RNAs Via Secm Ribosome Display 114
Kevin Baldridge, Amanda Bryant-Friedrich and Lydia Contreras

Engineering Secretion Machinery for High-Throughput Protein Production 117
Danielle Tullman-Ercek

Genome-Wide Search for Intracellular Factors Affecting RNA Folding Via in Vivo Oligonucleotide Hybridization 118
Kevin Vasquez, Jorge Vazquez-Anderson, Taylor Hatridge and Lydia M. Contreras

An Alternative Splicing Platform for Programming Protein Function 119
Melina Mathur and Christina D. Smolke

Transformable Facultative Thermophile Geobacillus Stearothermophilus NUB3621 As a Host Strain for Metabolic Engineering 120
Kristen Blanchard, Srebrenka Robic and Ichiro Matsumura

An Analytical Approach to Bistable Biological Circuit Discrimination Using Real Algebraic Geometry 121
Dan Siegal-Gaskins, Elisa Franco, Tiffany Zhou and Richard M. Murray

Characteristics of Protein Incorporated within Poly(vinyl alcohol) Hydrogel Membrane 122
Adrian Salgado

A Hybrid Riboswitch–Small RNA Molecule for Metabolic Engineering in an n-Butanol Pathway 123
Ashwin Lahiry, Samuel D. Stimple, David W. Wood and Richard A. Lease

Bacterial Inner Membrane Display for Isolating Intracellular Antibodies from a Naïve Library 125
Parisa Moghaddam-Taaheri and Amy J. Karlsson

Ensemble Modeling Identifies the Mechanism By Which Clpp Impacts NO• Defense Systems in E. coli 126
Mark P. Brynildsen and Jonathan L. Robinson

Re-Engineered β-Oxidation Reversal for the Synthesis of ω-Functionalized Products in Escherichia Coli 128
Seokjung Cheong

Medium-Chain Carboxylic Acid and Alcohol Production Via Engineered β -Oxidation in E. coli 129
Seohyoung Kim and Ramon Gonzalez

Cloning and Expression Profiling of Polycomb Gene, Vernalization Insensitive 3 from Tomato Solanum Lycopersicum L 130
Zainab Almutairi

Engineering, Predicting, and Understanding Cofactor Specificity in Ketol-Acid Reductoisomerases 131
Jackson K.B. Cahn, Sabine Brinkmann-Chen and Frances H. Arnold

A Click Chemistry Approach to Site-Specific Immobilization of a Small Laccase Enables Efficient Direct Electron Transfer in a Biocathode 132
Zhilei Chen and Dongli Guan

Rational Design of Antibody Cocktails to Treat Disease Caused By Bordetellae 133
Jennifer A Maynard, Annalee Nguyen, Ellen K Wagner, Edith Acquaye-Seedah, Roman Wolf, Eric Harvill and James Papin

Production of Fatty Acid-Derived Fuels and Chemicals in Saccharomyces Cerevisiae 134
Leo D. Espaux, Jay D. Keasling and Weerawat Runguphan

Improving Cell-Free Protein Synthesis through Genome Engineering of Escherichia coli Lacking Release Factor 1 135
Rey W. Martin, Seok Hoon Hong, Yong-Chan Kwon, Ben De Soye, Ioanna Ntai, Alexandra de Paz, Neil Kelleher and Michael C. Jewett

Constructing a Synthetic Metabolic Pathway Protects Escherichia coli Against the Acetic Acid Inhibition and Fulfills Enantiomerically Pure (R, R)-2,3-Butanediol Production 136
Xiaojun Ji and He Huang

A Reusable Immobilized Intein Segment That Eliminates Premature Cleaving and Thiol Requirements in Intein-Based Protein Purification 137
Changhua Shi, David Cain and David W. Wood

Design of Molecular Biosensors for New or Improved Polyketide Synthase Activities 138
Patrick C. Cirino and Zhiqing Wang

Modular Pathway Optimization for Production n-Butanol from Red Seaweed in Escherichia coli 139
Hyun Gyu Lim, Jae Hyung Lim and Gyoo Yeol Jung

Metabolic Engineering of Escherichia coli for Production of L-Tryptophan Using Synthetic Biology Tools 140
SungHo Jang, Sangwoo Seo and Gyoo Yeol Jung

Modular Ligand-Activated Enzymes from Antibody Mimetic Proteins 141
Nathan Nicholes, Amol Date, Pamphile Beaujean, Manu Kanwar, Pricila Hauk and Marc Ostermeier

Developing an Enzymatic Switch from HSV-Thymidine Kinase As a Potential Cancer Therapeutic 142
Nirav Shelat and Marc Ostermeier

Heterologous Reconstruction of a Synthetic Bacterial Oscillator 143
Anna H. Chen

Synthetic Biology Approach to Systematic Optimization of Protein Function 144
Sridhar Govindarajan

Design and Characterization of Synthetic, Minimal Promoters in Yeast 145
Heidi Redden and Hal Alper

Introduction of Heterologous Mevalonate Pathway in Methylobacterium Extorquens AM1 for High Production of Value-Added Compounds from Methanol 147
Wen-Liang Zhu, Chong Zhang, Lan-Yu Cui and Xin-Hui Xing

Antibody Mediated Protection in Whooping Cough Infection Due to Bordetella Pertussis 148
Edith Acquaye and Jennifer Maynard

New Biophysical Model for Characterization of Human Heart Valves 149
Juan Garcia, Jose Padilla, Gerardo Rivera-Silva and Maria Moreno

Reengineering the Genetic Code to Expand the Language of Biology Using a Cell-Free Synthetic Biology Approach 150
Bradley C. Bundy, Mark T. Smith, Jeremy Hunt, Amin Salehi, Andrew Broadbent and Matt Schinn

A Novel Storage Method for Umbilical Cord Blood 151
Gul Afshan and Eryn Hassemer

Design of Halotolerant Bovine Dnase Guided By Phylogenetic and Domain Structure Studies of Bacterial Dnases 152
Gediminas Alzbutas, Milda Kaniusaite, Algirdas Grybauskas and Arunas Lagunavicius

Composability and Design of Parts for Large-Scale Pathway Engineering in Yeast 153
Eric M. Young, Johannes A. Roubos, Ben Meijrink and Christopher A. Voigt

Cofactor Engineering for the Production of Short Chain Fatty Acid 154
Songi HAN and Ka Yiu San

Photocrosslinking with p-Azidophenylalanine Used to Identify the Location of Cic Binding on ERK 155
Alan Futran, Stanislav Y. Shvartsman and A. James Link

In-Depth Determination and Analysis of the Human Paired Heavy and Light Chain Antibody Repertoire 156
Brandon J. DeKosky, Takaaki Kojima, Alexa Rodin, Kam Hon Hoi, Jiwon Lee, Wissam Charab, Gregory C. Ippolito, Andrew D. Ellington and George Georgiou

Integrative Characterization of Human Cellular and Serological Antibody Repertoires Elicited by Seasonal Influenza Vaccine 157
Jiwon Lee, Daniel R. Boutz, Christopher Vollmers, Brandon J. DeKosky, Andrew P. Horton, Ellen M. Murrin, Daechan Park, Gregory C. Ippolito, Edward M. Marcotte, Stephen R. Quake and George Georgiou

Author Index

Multipurpose Bacterial Cell Factories: New Production Strains and Novel Tools

Bott, M., Forschungszentrum Juelich

A number of microorganisms have turned out to be particularly suitable for industrial production processes. With respect to amino acids, *Corynebacterium glutamicum*, a Gram-positive soil bacterium, is the preferred host for production in the million ton scale [1]. I will present newly developed *C. glutamicum* strains for production of amino acids and proteins, including genome-reduced strains, and novel tools for strain and enzyme development based on single-cell metabolite sensors, FACS screening, and recombineering.

[1] Eggeling, L. and Bott, M. (2005) Handbook of *Corynebacterium glutamicum*. CRC Press, Taylor & Francis Group, Boca Raton, Florida, USA

Non-Canonical Amino Acids as Probes of Protein Synthesis in Complex Biological Systems

Tirrell, D., California Institute of Technology

The genetic code, elucidated in the 1960s through the work of Nirenberg, Ochoa, and Khorana, provides a set of molecular instructions for turning DNA into proteins. Over the last decade, cells have been outfitted with modified molecular machinery that enables them to use non-standard (non-canonical) amino acids to make proteins. These developments are leading to new approaches to macromolecular design, protein evolution, biological imaging, and proteome-wide analysis of cellular processes. This lecture will emphasize the use of non-canonical amino acids to analyze protein synthesis in time-resolved and cell-selective fashion in complex biological systems including live animals.

Signals from the Surface to Promote Human Pluripotent Stem Cell Differentiation

Kiessling, L., University of Wisconsin- Madison

Human pluripotent stem cells (human embryonic stem cells and induced pluripotent stem cells) have the ability to self-renew indefinitely and differentiate into specialized types. Soluble signals (e.g., growth factors or small molecules) have been identified that promote human pluripotent stem (hPS) cell differentiation. Most standard differentiation protocols employ Matrigel, an undefined and complex mixture of over 1800 proteins. The role of the insoluble signals in directing differentiation is largely unexplored. To mine this molecular space, we devised arrays of chemically defined surfaces composed of self-assembled monolayers of peptides. Using the results from our array screen, we generated peptide-modified surfaces and used them to elucidate the molecular mechanisms by which insoluble signals guide hPS cell differentiation to each of the primary germ layers. We identified surfaces with surprising assets: They not only permit human pluripotent stem cells to differentiate to specific lineages but even *instruct* them to do so. Our findings highlight the dramatic effects of insoluble cues on stem cell pluripotency and differentiation. They also underscore the utility of defined surfaces to dissect critical signaling pathways.

Development of Microbial Cell Factories for Biorefineries

Kondo, A., Graduate School of Engineering, Kobe University

To build an energy and material secure future, we must pioneer the next generation of renewable fuels and chemicals using environmentally benign production processes. Since biomass represents an abundant carbon-neutral renewable resource for the production of biofuels, numerous environmental and social benefits could result from the replacement of petroleum-based transport fuels with bioethanol converted from biomass. One of the key technologies for the development of biorefineries is cell surface engineering, which is a powerful tool for engineering and functionalizing many organisms. Using the technology, various kinds of functional proteins, such as enzymes, can be expressed on the cell surface without loss of cell activity. The display of amylolytic and cellulolytic enzymes on the surface of Saccharomyces cerevisiae has accomplished direct ethanol production from starchy and cellulosic biomass. Moreover, the display of hemicellulase on the surface of S. cerevisiae that has a xylose-assimilating metabolic pathway has enabled production of ethanol from hemicellulosic materials. Furthermore, reutilization of the cell-surface engineered yeast has an advantage in the reduction of enzyme cost, which enables reuse of enzymes on the cell surface by collecting the cells. Thus, cell surface engineering is a promising technology for the development of a consolidated bioprocess by integrating enzyme production, saccharification and fermentation. Regardless of the biomass hydrolysis, metabolic engineering of microorganisms is emphasized for the efficient production of ethanol from biomass.

Reinforcing Synthetic Biology Against Evolutionary Failure

Barrick, J. E., University of Texas at Austin

Originally presented on: 1/12/2015 0:00:00 - 0:25:00

Unwanted or unintended evolution can rapidly degrade the performance of complex DNA-encoded devices and complicate large-scale strain engineering efforts. *Acinetobacter baylyi* ADP1 has been proposed as a new host organism for synthetic biology and metabolic engineering because it is naturally transformable. To understand the major causes of genetic instability in this "undomesticated" bacterium, we performed laboratory evolution experiments with ADP1 cultures under conditions where either adaptive mutations or genetic drift dominated. By analyzing whole-genome sequencing data from evolved strains using the *breseq* software program developed by our research group, we discovered that IS*1236* transposable elements were largely responsible for the loss of several key traits during laboratory propagation, including competence for DNA uptake. To stabilize the ADP1 genome, we engineered an ISx strain in which all copies of IS*1236* — the only native transposable element in ADP1 — were deleted from the chromosome. The ISx ADP1 strain exhibits a greater reduction in overall spontaneous mutation rates than predicted from the incidence of IS*1236* insertions and an unexpected increase in transformation efficiency. These results may be due to the decreased accumulation of extracellular DNA from lysed cells in ISx compared to wild-type ADP1 cultures. Finally, we have demonstrated a generalizable directed evolution approach for discovering mutations that increase the evolutionary lifetime of a DNA device encoding a costly engineered function. Using this strategy, we isolated *E. coli* variants that are able to maintain fluorescent protein expression from a plasmid for more than twice as long as wild-type hosts. Genome sequencing of these variants found that they have mutations that likely result in lower mutation rates that would stabilize maintenance of any DNA sequence in these hosts. Together, these studies show that it is possible to engineer and evolve more genetically reliable host organisms to effectively prevent evolutionary failure in synthetic biology.

Engineering Smarter and Stronger T Cells for Cancer Immunotherapy

Chen, Y., University of California, Los Angeles

Originally presented on: 1/12/2015 0:50:00 - 1:15:00

Adoptive T cell therapy for cancer has demonstrated exciting potential in treating relapsing cancers. In particular, T cells that express synthetic chimeric antigen receptors (CARs) specific for the B-cell marker CD19 have shown impressive results in clinical trials for various B-cell malignancies, prompting avid interest from both scientific and entrepreneurial communities in recent years. However, CD19 CAR-T cell therapy remains the only robustly effective T-cell immunotherapy to date, and several obstacles remain to be overcome before the full potential of adoptive T-cell therapy can be realized. Here, we present several strategies for the engineering of T cells with stronger anti-tumor functions and greater robustness against evasive mechanisms employed by cancer cells. We discuss the design, construction, and implementation of multi-input CARs to increase tumor specificity and decrease the probability of mutational escape by tumor cells. We further present the design of synthetic circuits to reroute signaling pathways triggered by tumor-secreted cytokines, thus negating the immunosuppressive effects of the tumor microenvironment. Finally, we present an engineered cytotoxic protein that triggers target-cell death upon recognition of an intracellular oncoproteins, thus expanding the repertoire of detectable tumor markers beyond surface-bound antigens. These strategies combine to address critical limitations facing adoptive T-cell therapy, providing potential treatment options for diseases that are otherwise incurable with current technology.

Synthetic Regulators for Evolutionary Metabolic Engineering

Jung, G. Y., POSTECH

Originally presented on: 1/12/2015 1:15:00 - 1:40:00

Metabolic Engineering aims to purposeful design or redesign biological system for the production of commercially valuable chemicals such as biofuels, platform chemicals and biologically active compounds. To achieve the successful design or redesign of the biological systems, robustness of naturally occurring biological systems has to be relieved so that cells can be easily redesigned. Although extremely huge efforts have been made to find genetic target to improve metabolic function of the microorganisms, there still exists the additional room for the non-rational approach. One of the most critical problems of metabolic engineering is especially robustness of the biological systems. Bacterial cells are generally evolved at the various levels from DNA to protein for maintaining their robustness against the changing circumstances. Therefore, general strategy to modify cellular physiology depending the robustness or flexibility of the biological systems should be required. In this study, we developed intracellular metabolite sensor named "riboselector" to regulate metabolic distribution will be presented. The potentials of the platform technology developed in this study for the application to the production of biofuels and commodity chemicals.

Synthetic Biology with a Cell-Free TX-TL System: Metabolism, Gene Circuits and Minimal Cell in a Test Tube

Shin, J., MIT
Caschera, F., University of Minnesota
Noireaux, V., University of Minnesota
Originally presented on: 1/12/2015 1:40:00 - 2:05:00

Cell-free transcription-translation (TX-TL) systems were developed in the 60s to study the process of protein synthesis and gene regulation in bacteria. With the invention of highly efficient cell-free expression systems in the 90s, in vitro protein synthesis became used for biotechnological applications and in the industry as an alternative to the recombinant protein technology. Over the past decade, cell-free TX-TL systems have become promising platforms for in vitro synthetic biology and bioengineering.

I will present a cell-free TX-TL toolbox specifically developed to construct in vitro biochemical systems programmed with DNA. Our original idea was to reconstruct elementary gene circuits in isolation free of endogenous genetic material. Recently we showed that our custom-made TX-TL system can execute large DNA programs. The bacteriophage T7, composed of about 60 genes, is entirely synthesized in a single test tube reaction from its genomic DNA. Replication of the T7 DNA instructions occurs concurrently with phage expression and self-assembly. Encapsulated inside cell-sized liposomes, the TX-TL system is used to construct a prototype of minimal cell. I will present this cell-free TX-TL platform, our last experiments and how this system is used to engineer gene circuits and self-organized entities as well as how we can address biophysical questions.

Engineering alcohol tolerance in yeast fermentations

Stephanopoulos, G., Massachusetts Institute of Technology
Fink, G., Whitehead Institute
Lam, F. H., Massachusetts Institute of Technology

Originally presented on: 1/12/2015 0:00:00 - 0:25:00

Ethanol toxicity in yeast *Saccharomyces cerevisiae* limits titer and productivity in industrial production of transportation bioethanol. We show that strengthening the opposing potassium and proton electrochemical membrane gradients is a mechanism that enhances general resistance to ethanol. The same mechanism and protection method is also applicable to higher alcohols. Elevation of extracellular potassium and pH physically bolster these gradients, increasing tolerance to higher alcohols and ethanol fermentation in commercial and laboratory strains (including a xylose-fermenting strain) under industrial-like conditions. A key finding is that specific productivity (production per *viable* cell) remains largely unchanged with improvements deriving from heightened population viability. As such, yeast cultures remain highly (and equally) productive for as long as they stay viable. Likewise, up-regulation of the potassium and proton pumps in the laboratory strain enhances performance to levels exceeding industrial strains. Although genetically complex, alcohol tolerance can thus be dominated by a single cellular process, one controlled by a major physicochemical component but amenable to biological augmentation.

The Quest for Synthetic Methylotrophy

Papoutsakis, E. T., University of Delaware
Whitaker, W. B.,
Sandoval, N. R., University of Delaware
Bennett, R. K.,
Fast, A. G., University of Delaware
Originally presented on: 1/12/2015 0:25:00 - 0:50:00

Microbial systems are attractive platforms for producing biofuels and commodity chemicals from natural gas (methane) and methanol. While organisms that naturally utilize methane and/or methanol have been known for years and have been explored for production of single-cell protein and few large molecules (PHB and carotenoids), no reports exist for the industrial production of small metabolites. Synthetic methylotrophy refers to the development of recombinant microbial strains that can utilize reduced one-carbon (C1) chemicals, and notably methane, methanol and formaldehyde as sole carbon and energy source and/or as co-substrates to sugars and other current substrates of microbial cultures. The goal would be to use these C1 substrates to produce useful metabolites such as biofuels, commodity chemicals, chemical precursors and biopolymers. The impetus derives from two reasons. The first is that methane from natural gas constitutes an enormous in size resource for generating energy and producing chemicals. In the last few years, the availabilities of methane and its oxidation product methanol have been increasing at dramatic rates while their prices have been decreasing. The second reason is that methane and methanol are considerably more reduced than most sugar substrates and thus have to potential to enable the biological production of biofuels and chemicals with significantly increased yields. We will discuss the native biology of methylotrophy and how it can be engaged to develop superior methylotrophic strains using synthetic biology and metabolic engineering, with emphasis on the use of methanol and formaldehyde. Engineering platform organisms to utilize methanol for producing small metabolites is recently actively pursued, for now largely based on the patent literature. Metabolic engineering and synthetic biology of platform organisms, and notably *E. coli*, are key for technology advances to enable the conversion of the overabundant and inexpensive methane and methanol into useful chemicals and fuels.

Metabolic Engineering of Saccharomyces Cerevisiae for the Synthesis of Polyketides and Fatty Acids

Da Silva, N. A., University of California, Irvine
Leber, C., University of California, Irvine
Cardenas, J., University of California, Irvine
Choi, J. W., University of California, Irvine
Fernandez-Moya, R., University of California
Originally presented on: 1/12/2015 0:50:00 - 1:15:00

Metabolic Engineering of *Saccharomyces cerevisiae* for the Synthesis of Polyketides

and Fatty Acids

Nancy A. Da Silva, Christopher Leber, Javier Cardenas, Jin Wook Choi, Ruben Fernandez-Moya

Department of Chemical Engineering and Materials Science, University of California, Irvine

Polyketides and fatty acids are of critical importance as biorenewable chemical precursors, biofuels, and pharmaceuticals. Both are synthesized via complex polyketide or fatty acid synthases, with many using acetyl-CoA and malonyl-CoA as starter and extender units. We have engineered the yeast *Saccharomyces cerevisiae* for the production of these valuable compounds and to allow the synthesis of novel variants. We have combined enzyme engineering (of the pathway and synthase enzymes), extensive metabolic pathway engineering for increased cofactor and precursor pools, and cultivation strategies to substantially increase titers and yields of a variety of products, including 6-methylsalicylic acid (6-MSA), dihydromonocolin L (DML; precursor to lovastatin), fatty acids (FAs) of varying lengths, and triacetic acid lactone (TAL). Fatty acids and triacetic acid lactone are attractive biomolecules due to their straightforward conversion into a broad range of biofuel and biochemical products. TAL can be converted via chemical catalysis to sorbic acid, 1,3-pentadiene, and a variety of other end products. *S. cerevisiae* was engineered for the high-level production of TAL by overexpression of the *Gerbera hybrida* 2-pyrone synthase (2-PS), engineering of the yeast metabolic pathways, and implementation of various cultivation strategies. These interventions increased both yield and titer 40 to 150-fold (>5 g/L TAL), levels far higher than previously reported. Fatty acids are also of interest as both biofuel and chemical precursors. We have introduced heterologous fatty acid synthases into *S. cerevisiae* to allow the synthesis of short (C6-C8) and medium (C10-C12) chain free fatty acids, and have done extensive pathway engineering to increase the levels and secretion of these and long-chain free fatty acids (C16-C18) to the culture medium. As an example, *S. cerevisiae* was engineered to overproduce free fatty acids by increasing carbon flux from glucose into the fatty acid and neutral lipid forming pathways, and by preventing the degradation and reactivation of these fatty acids. A unique combination of six gene knockouts and overexpression of two genes resulted in extracellular FFAs at a titer of 2.2 g/L, 4.2-fold greater than the highest level of extracellular free fatty acids reported for *S. cerevisiae*. In the presentation, we will discuss the critical pathways engineered, examine the synergy between successful strategies for the various polyketide products, and quantify the improvements obtained.

Exploring P450 Expression in Escherichia coli for the Synthesis of Complex Molecules

Biggs, B.,
Lim, C. G., University at Buffalo, the State University of New York
Das, A.,
De Mey, M.,
Parayil, A., Massachusetts Institute of Technology
Originally presented on: 1/12/2015 1:15:00 - 1:40:00

Biocatalysis projects to play a significant role in the future of chemical synthesis, especially for the creation of complex natural products, their analogs, and new chemical entities built upon their scaffolds. This is due to the ability of enzymes to leverage the incredible precision of a three dimensional binding pocket to perform highly specific functionalizations, overcoming a molecule's inherent reactivity biases. Added within the context of metabolic engineering, enzymatic biocatalysis possesses tremendous opportunity to improve scalability, minimize environmental impact, and reduce costs by employing fast growing organisms, performing reactions at near neutral conditions, and converting affordable feedstock to compounds of interest.

Of the potential chemical targets, terpenoids, the largest class of plant specialized metabolites, are particularly interesting as they hold relevance to flavor, fragrance, pharmaceutical, and fuel targets. Realizing the full impact of terpenoids, though, will hinge upon our ability to efficiently perform complex functionalizations of terpene backbone molecules. To this end, harnessing the capabilities of the P450 family of enzymes will be essential, as they represent nature's oxygenase of choice for functionalizing unreactive carbons with specificity. Herein we present our work aimed at obtaining high titers of functionalized terpenoid molecules in the Taxol biochemical pathway. We will focus on functional expression of plant P450s in *E. coli*, balancing metabolic pathway flux, improving catalytic efficiency by modulating reductase partners (CPR), and performing detailed characterization of P450 enzymes with *in vitro* assays, all in order to construct high-performing systems.

Co-Evolution of Affinity and Stability for Domain Antibodies That Recognize Hydrophobic Antigens

Tessier, P. M., Rensselaer Polytechnic Institute
Originally presented on: 1/12/2015 0:00:00 - 0:25:00

Phage and yeast surface display are invaluable methods for evolving and tailoring the properties of biomolecules ranging from peptides and small proteins to multidomain enzymes and full-length antibodies. These methods have been used in many innovative ways for improving not only biomolecular activity but also stability as well. We have sought to use yeast surface display to evolve the affinity and stability of single-domain (V_H and V_L) antibodies specific for hydrophobic peptides and proteins linked to Alzheimer's and Parkinson's diseases. Interestingly, we find that the binding affinity of selected antibody mutants is anti-correlated with folding stability. Multiple rounds of mutagenesis and selection result in high-affinity domain antibodies that are unfolded on the surface of yeast and as isolated domains when produced in bacteria. The unfolded antibodies are soluble and well-expressed in bacteria, suggesting that the yeast secretory pathway removes aggregation-prone variants. To improve selection of stable domain antibodies, we have replaced sequence-specific antibodies used for detecting yeast display levels with conformational probes specific for folded antibodies. This leads to significant improvements in co-evolution of antibody affinity and stability. We are currently employing this approach for optimizing domain antibodies specific for soluble and aggregated forms of several hydrophobic (amyloid-forming) proteins, as well as evaluating these evolved antibodies for their ability to detect and inhibit toxic protein aggregation.

Engineering Proteases to Detect Post-Translational Modifications

Varadarajan, N., The University of Houston

Originally presented on: 1/12/2015 0:25:00 - 0:50:00

Proteases, enzymes that catalyze the hydrolysis of amide bonds in peptides and proteins, are ubiquitous across all forms of life and affect every protein in its natural life cycle. Engineering programmed specificity onto stable protease scaffolds holds promise as a route to a new generation of useful proteases with requisite biotechnological properties. One of the long-standing applications of proteases has been in generating peptides for mass spectrometry (MS) analyses in proteomic experiments. Since post-translational modifications (PTMs) can be labile, proteomic techniques relying on tandem MS can have difficulty in achieving comprehensive coverage of PTM sites within proteins. Proteases which selectively recognize PTMs may improve their identification in complex proteome samples and are easily adaptable to current proteomic methodologies. Chymotrypsin, trypsin and elastase are a family of serine proteases that have superimposable tertiary structures but recognize chemically differing substrates. Trypsin cleaves after basic residues (Arg/Lys) while chymotrypsin cleaves after large hydrophobic residues (Phe/Tyr/Trp), leading biochemists to believe that since the backbone architecture was identical, the origin of the differing substrate recognition capabilities must arise from the chemistry of the actual side-chains that comprise the substrate binding pockets. Unfortunately however, site-directed mutational swapping of selected residues within the binding pocket of trypsin did not endow chymotrypsin like reactivity and required the grafting of extended loops outside the substrate binding pocket to bring about the change in substrate specificity. These and other similar efforts, while noteworthy, serve to highlight the challenges associated with molecular redesign of substrate binding specificity within proteases.

In the current work, we have utilized a combinatorial approach to engineer the substrate binding pocket of chymotrypsin, and isolate variants that can recognize the presence/absence of glycosylation and phosphorylation modifications. This required us to develop a new platform to display mammalian chymotrypsin B on the surface of *E.coli* by fusing it with C-terminal barrel domain of antigen 43, a native membrane protein. Eleven residues in substrate binding pocket, chosen based on sequence alignment of chymotrypsin-fold proteases were randomized using oligonucleotides with NNS codons, cloned and transformed into *E.coli* MC1061 to obtain a library of 10^7 unique variants. High-throughput screening of the library was performed using FACS for simultaneous selection for activity towards Asn, and counter-selection against wild type-like activity, using peptide substrates containing a proteolysis-sensitive linker sandwiched between orthogonal FRET pairs. After six rounds of sorting, we observed sequence convergence from a library of 10^7 to a ChyB-Asn variant that showed desired fluorescence profile when measured using flow cytometer. Solution phase kinetics was employed to measure the Michaelis-Menten parameters of this variant and mass spectrometry used to confirm proteolysis after Asn containing peptide sequences. We anticipate that that our results can be a stepping stone for engineering differing substrate specificities onto the chymotrypsin scaffold and can help design the rules for molecular recognition within proteases. In parallel, we are also investigating the utility of the Chyb-Asn variant for glycoproteomic studies. Finally, our surface-dsiplay assisted combinatorial engineering methodology can readily be extended to engineering specificity towards other PTMs and we have also identified a chymotrypsin variant showing specific activity towards pTyr peptide substrates

Computational Strain Design for Improved Productivity, Yield and Robustness

Mahadevan, R., University of Toronto

Originally presented on: 1/13/2015 0:00:00 - 0:25:00

Bioprocess development for biofuels and biochemicals typically requires several rounds of metabolic engineering to meet process targets including product yield, titer and productivity, all of which impact the process economics. Similar advances in computational modeling techniques have allowed the development of genome-scale models of metabolism in several organisms and their use in rational design of strains. In the first part, a rational approach based on bi-level optimization to enhance bioprocess productivity by forcing co-utilization of substrates will be shown. Experimental results from the application of this approach to enforce substrate co-utilization in *Escherichia coli* will be discussed. In the next part of the talk, a novel nested nonlinear optimization method for metabolic engineering resulting in over hundred different strain design strategies for biochemicals production will be presented. We will also examine the role of the redundant pathways from a design perspective and present computational results on how these pathways is valuable for robust design.

Machine-Based Learning for Rational Engineering of Biomolecular Libraries

Contreras, L. M., The University of Texas at Austin
Vasquez, J.,
Reyes, K.,
Powell, W., Princeton University
Originally presented on: 1/13/2015 0:50:00 - 1:15:00

Ribonucleic acids (RNAs) have been become highly relevant to our understanding of cellular regulation. Although in highly diverse ways, RNAs effect cellular changes by altering recognition patterns with their targets.. Recent work in our laboratory has focused on the discovery of non-coding RNAs of relevance to various bacterial stress responses. An important part of characterizing these RNAs has been understanding their structural rearrangements in the native context of the cellular milieu. Although RNA structure has been extensively studied using a variety of *in vitro* and *in silico* techniques, to date, only a few RNA structural characterization techniques have been used intracellularly. In this work, we are taking advantage of principles derived from synthetic biology and from rational biomolecular engineering to design fluorescence-based libraries of probes capable of in vivo detection of structural perturbations in target RNAs. While we have been experimentally successful in using flow cytometry data to perceive structural differences between RNA conformers, a major challenge has been the co-engineering of optimal *probe libraries* that, in an interdependent and combinatorial way, provide the most structural information. We will discuss the application of machine-based learning strategies for automating the engineering of these larger molecular libraries that are coupled by functionality. These strategies have been largely successful in the rational design of non-biological materials. Our preliminary results demonstrate that a strong connection between machine learning and molecular engineering could be designed based on imposing mechanistic descriptions of the functionality of the library. We will also discuss potential implications of this approach for engineering molecules in a way that optimizes multiple traits.

De Novo Biosynthesis of Terminal Alkyne-Tagged Natural Products and Applications

Zhang, W., University of California, Berkeley

Originally presented on: 1/13/2015 1:15:00 - 1:40:00

Natural products are important small molecules widely used as drugs, pesticides, herbicides, and biological probes. Tagging natural products with a unique chemical handle, such as an alkyne functionality, enables the visualization, enrichment, quantification, and mode of action study of natural products through bioorthogonal chemistry. However, the labeling of natural products has not been adequately explored compared to that of macromolecules. We recently developed a new general platform to tag natural products by characterizing and engineering the novel terminal alkyne synthetic biology that living systems offer. In addition to directly producing orthogonally functional natural product analogues that can be subjected to facile chemical modification for drug screening, our work may significantly advance the field of natural product research through coupling with diverse azido analytical handles. For example, using a fluorogenic probe, in situ detection and quantification of natural products in producing cell cultures can be accomplished for the first time, enabling the facile selection of the best overproduction strain. Furthermore, since azide-alkyne click chemistry has recently emerged as one of the most powerful tools in drug discovery and chemical biology, our findings open the door to numerous biological applications in which in situ enzymatic generation of a terminal alkyne is required or preferred.

Development and Analysis of Precursor Production Strains for Chemical Production

Zhang, X., University of Wisconsin Madison
Tervo, C. J., University of Wisconsin - Madison
Reed, J., University of Wisconsin-Madison
Originally presented on: 1/13/2015 1:40:00 - 2:05:00

Microbes have been engineered to produce a variety of chemicals, including biofuels, commodity chemicals, and therapeutics. Production can be enhanced by connecting biosynthesis pathways to host metabolism and optimizing the pathway's expression. However, another important step in engineering microbes to produce desired chemicals is to increase production of their chemical precursors. In this work, we evaluated what precursors are needed to synthesize a variety of chemicals, and then designed and engineered strains to produce one of these precursors—pyruvate.

We systematically evaluated what potential non-native products could be produced by *Escherichia coli* (using both native and heterologous pathways) and then evaluated their distance (in terms of metabolic reactions steps) from central metabolic precursors. Using a genome-scale metabolic model of *E. coli* and a set of potential heterologous reactions (from the KEGG database), ~1,800 non-native products could potentially be produced in *E. coli* using heterologous enzymes. Of these 284 have reported commercial applications. The set of 284 chemicals were subsequently analyzed to identify how many heterologous reactions were needed to enable their production and whether they were within five reaction steps of a central metabolic precursor. This analysis identified that of the six central metabolic precursors considered, pyruvate was the closest precursor to the most non-native valuable products.

Since pyruvate has industrial applications and is within 5 reactions steps of many valuable non-native products, we sought to develop a strain of *E. coli* that could produce pyruvate at high yields. A high-yield pyruvate producing strain has great potential for creating additional strains to produce a variety of other valuable chemicals. Guided by a genome-scale metabolic model of *Escherichia coli*, we identified different strategies for enhancing production of pyruvate from glucose. The targeted gene deletions minimize acetyl-CoA production, undesired product (acetate and lactate) formation, and NAD(P)H formation. We constructed a number of strains and five of them achieved yields of more than 0.9 g pyruvate per g substrate (94% theoretical yield) under aerobic conditions.

Further genetic modifications were made to evaluate whether additional products (derived from pyruvate) could also be made using these strains. To produce ethanol, pyruvate formate-lyase (PflB) was deleted, and pyruvate decarboxylase (Pdc) and alcohol dehydrogenase II (AdhB) from *Zymomonas mobilis* were over-expressed in the engineered pyruvate strains. These genetically modified strains fermented glucose to ethanol with up to yields of 0.43 g ethanol per g substrate (~84% of theoretical yield).

These results illustrate how computational models can be used to prioritize precursor-based strategies and identify genetic modifications to enhance precursor production. We also showed that precursor over-production strains can be further modified to produce other valuable products. Current efforts are focused on engineering the pyruvate production strains to produce other native and non-native products, as well as developing other precursor over-production strains.

Micromachines and motility: Single and collective cancer cell invasion

Zaman, M. H., Boston University

Originally presented on: 1/13/2015 0:00:00 - 0:25:00

Cell migration and cell-matrix interactions regulate some of the most critical stages of tumor invasion and metastasis. However, our understanding of the systems level behavior of these processes is often plagued by culturing cells in artificial environments, focusing on a single protein or ignoring the interplay between cell-signaling and cellular mechanics. To understand the bi-directional interaction of cellular machinery and cell-signaling in single and collective cellular motion, we have developed an integrated experimental and computational approach combining mitochondria-tracking microrheology, microfluidics, and Brownian dynamics simulations. Our work analyzes, in both single cells and in cellular clusters that are up to a millimeter in diameter, in 2D and in 3D, both passive thermal and active motor-driven processes within the cell and demonstrate how active internal fluctuations are modulated in native like environments. Our results demonstrate the synergistic roles of tension and signaling, dimensionality and disease progression and show that invasive cells exhibit more solid-like internal motions in 3D compared to 2D. In addition, cells in 2D are more sensitive to actin disruption than in 3D. Our integrated approach has resulted in discovery of novel mechanisms of collective cellular motion, new regulators of cellular persistence in 3D environments and the role of ECM in modulating chemoresistance in breast and prostate cancer.

Multiscale Models of Antibiotic Cellbots

Geldart, K., University of Minnesota
Forkus, B., University of Minnesota
Kyriakou, P., University of Minnesota
Kaznessis, Y. N., University of Minnesota
Vlysidis, M., University of Minnesota
Originally presented on: 1/13/2015 0:25:00 - 0:50:00

Foodborne gastrointestinal infections are significant causes of morbidity and mortality worldwide. Alarmingly, because of the extensive, non-therapeutic use of antibiotics in agriculture, foodborne bacteria are emerging that are resistant to our most potent drugs.

We will discuss a novel approach to reduce the use of antibiotics in food-producing animals and to treat gastrointestinal infections. We engineer lactic acid bacteria (LAB) that express and release antimicrobial peptides (AMPs). LAB are part of the gastrointestinal microflora and can be safely delivered with known benefits to humans and animals. AMPs are proteins that can be readily produced by LAB. One unique aspect of our approach is the use of synthetic promoters that precisely regulate the delivery of AMP molecules.

At the heart of proposed efforts are multiscale models that guide explanations and predictions of the antagonistic activity of recombinant LAB against pathogenic strains. Models are developed to quantify how AMPs kill bacteria at distinct but tied scales. Using atomistic simulations the various interaction steps between peptides and cell membranes are explored. Mesoscopic models are developed to study ion transport and depolarization of membranes treated with AMPs. Stochastic kinetic models are developed to quantify the strength of synthetic promoters and AMPs expression. We will also present a closure scheme for chemical master equations, providing a solution to a problem that has remained open for over seventy years.1

Experimentally, we engineer lactic acid bacteria to inducibly produce antimicrobial peptides. We have built a library of synthetic biological constructs.2,3 We test these modified bacteria against pathogenic bacteria. We will present results against salmonella and enterococcus.4,5

1. Smadbeck P, Kaznessis YN. A closure scheme for chemical master equations. Proc Natl Acad Sci U S
A. 2013 Aug 27;110(35):14261-5. doi: 10.1073/pnas.1306481110.
2. Ramalingam KI, Tomshine JR, Maynard JA, Kaznessis YN. Forward engineering of synthetic bio-logical
AND gates, Biochemical Engineering, 47, 38-47, 2009.
3. Volzing K, Billiouris K, Kaznessis YN. proTeOn and proTeOff, New Protein Devices That Inducibly
Activate Bacterial Gene Expression, ACS Chemical Biology, 6, 1107-1116, 2011.
4. Volzing K, Borrero J, Sadowsky MJ, Kaznessis YN. Antimicrobial Peptides Targeting Gram-negative

Pathogens, Produced and Delivered by Lactic Acid Bacteria. ACS Synth Biol. 2013, 15;2(11):643-50.
doi: 10.1021/sb4000367
5. Borrero J, Chen Y, Dunny GM, Kaznessis YN. Modified Lactic Acid Bacteria Detect and Inhibit
Multiresistant Enterococci.ACS Synth Biol. 2014, In press

Nanoscale Architecture of Tension Generation within Focal Adhesions

Dunn, A., Stanford University

Originally presented on: 1/13/2015 0:50:00 - 1:15:00

Understanding the molecular mechanisms by which cells respond to physical cues such as extracellular matrix (ECM) stiffness and ligand density constitutes a critical bottleneck in the field of tissue engineering. Cells adhere to the ECM via integrins, a class of heterodimeric, transmembrane proteins that physically link the cell cytoskeleton to ECM proteins. The cytoplasmic domains of integrins in turn recruit hundreds of proteins that collectively comprise focal adhesions (FAs), highly complex, dynamic assemblies that mediate "outside-in" mechanosensing and "inside-out" traction force generation. How these assemblies mediate force sensing and force transmission is thus the subject of great interest. Current techniques lack the spatial resolution to directly visualize mechanical forces within individual FAs, posing a critical technical limitation in addressing this general question. We engineered Förster resonance energy transfer (FRET)-based molecular tension sensors (MTSs) to directly visualize the mechanical force exerted by individual integrins in living cells. Simultaneous super-resolution light imaging of MTSs and GFP-tagged cellular proteins results in maps of the force-producing structures within FAs with <100 nm spatial resolution, allowing us to, for the first time, directly correlate local force generation with the presence of specific integrins and cytoskeletal proteins. We find that $\alpha_v\beta_3$ integrin localizes to high force regions, whereas $\alpha_5\beta_1$ integrin localization is more diffuse. The canonical FA proteins paxillin, vinculin, talin, and α-actinin colocalize with force production to varying degrees. Surprisingly, paxillin, which is not generally considered to play a direct role in force transmission, shows a higher degree of spatial correlation with force than vinculin, talin, or α-actinin, proteins with hypothesized roles in mechanotransduction. In addition, simultaneous time-lapse imaging of either GFP-paxillin or GFP-α-actinin with MTS-measured local traction forces reveals that paxillin and tension are closely related in both space and time in assembling and disassembling adhesions, while α-actinin exhibits a more complex relationship with tension. The high degree of spatial correlation of both paxillin and $\alpha_v\beta_3$ integrin with mechanical tension suggests that these proteins may play direct roles in cellular mechanotransduction.

Automated Physics-Based Design of Synthetic Riboswitches from Diverse Aptamers

Salis, H. M., Pennsylvania State University

Originally presented on: 1/13/2015 1:15:00 - 1:40:00

Riboswitches are RNA-based sensors that use an aptamer domain to bind ligand, change shape, and alter gene expression levels; typically, by modulating the translation or transcriptional termination of mRNAs. While natural riboswitches have evolved as exquisite sensors, it remains difficult to engineer non-natural riboswitches that utilize different aptamers to detect and respond to ligands of interest, even though their application as biosensors, medical diagnostics, and metabolic controllers would be transformative. Previously, we developed a statistical thermodynamic model of ribosome-mRNA interactions that predicts translation rate from mRNA sequence. By incorporating the thermodynamics of ligand-binding, RNA folding, structural switching, and macromolecular crowding interactions, we developed a sequence-structure-function relationship for translation-regulating riboswitches formulated entirely using physical chemistry calculations. The model is combined with an optimization algorithm to enable the automated design of synthetic riboswitches using any selected aptamer domain.

Overall, we applied automated design to generate over 60 synthetic riboswitches using several different aptamers, which were characterized inside cells and within *in vitro* conditions. Using this data, we validated model predictions under equilibrium, ligand-limited, and non-equilibrium conditions; and in dilute or crowded environments. The sequence-structure-function model explained how changing aptamer structure and affinity, co-transcriptional ligand binding, ligand and mRNA concentrations, and the surrounding mRNA sequences controlled riboswitch activation. Our algorithm reliably generated riboswitches with high activation ratios, up to 383-fold, enabling non-experts to develop chemical sensors for their own applications. Finally, using a so-called "perfect riboswitch", we quantify the outer limits of biological sensing and highlight the potential applications of riboswitch-based sensing.

Zymergen

Serber, Z., Zymergen

Originally presented on: 1/13/2015 0:00:00 - 0:25:00

With unparalleled scale in laboratory automation and computation, Zymergen builds and optimizes the microbes that serve as cellular factories in the $100B+ industrial biotech sector. Zymergen's products -- optimized cellular factories -- drop directly into our customers' existing bio-manufacturing workflows for immediate impact on their bottom line without additional investments in infrastructure and capital equipment.

Zymergen employs radical empiricism to design, generate, and optimize microbes by rewriting their DNA. Our approach combines biology, robotic automation, and proprietary computational and analytic methods to industrialize what to date has been a slow, risky, essentially artisanal process. As a result, we generate microbes that produce novel chemicals, advanced materials, and pharmaceuticals faster and at lower costs than competing technology.

Engineering Microbial Fatty Acid Biosynthesis as an Industrial Biotechnology Platform Steve Del Cardayre*

del Cardayre, S.,

Originally presented on: 1/13/2015 0:25:00 - 0:50:00

REG Life Sciences acquired LS9 in January of 2014 and is advancing its microbial fatty acid technology platform. Microbial fatty acid metabolism has been demonstrated as a versatile platform for the production of diverse renewable chemicals and fuels. From novel raw materials for the oleochemical and food industries to drop-in and advantaged products for the fuel, commodity, and high-value chemical industries, new applications of this pathway continue to emerge. REG Life Sciences is focused on the systematic control of microbial fatty acid metabolism to synthesize new chemical families, control structure within each family, and optimize production efficiency to support scalable commercial processes. This talk shall briefly review the REG Life Sciences technology platform and highlight specific examples of new product pathways and the optimization of our most advanced product platforms.

Orthogonal Genetics for Rapid Evolution and Synthetic Biology

Chang C. Liu
Assistant Professor
Department of Biomedical Engineering
Department of Chemistry
University of California, Irvine

January 13, 2015

Orthogonal Replication

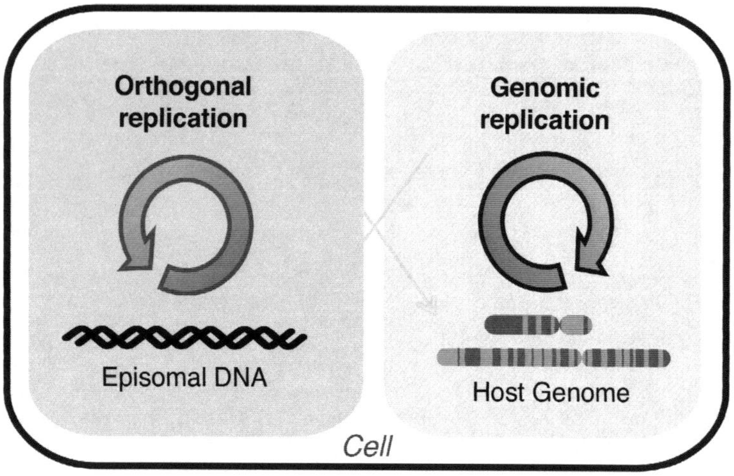

Orthogonal Replication

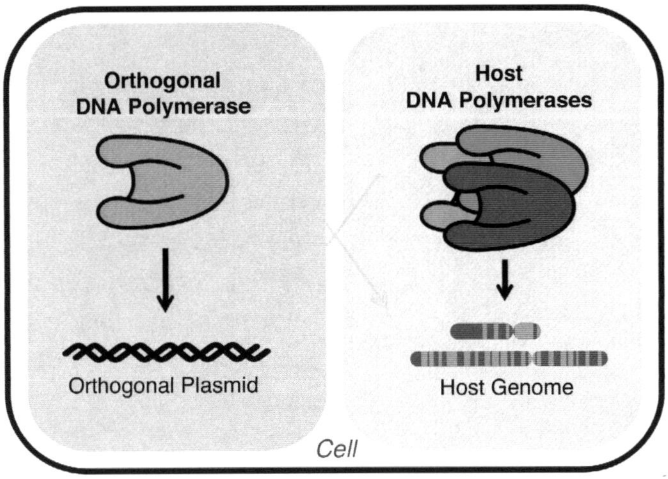

An independent DNA polymerase/DNA plasmid pair

Orthogonal Replication: Why?

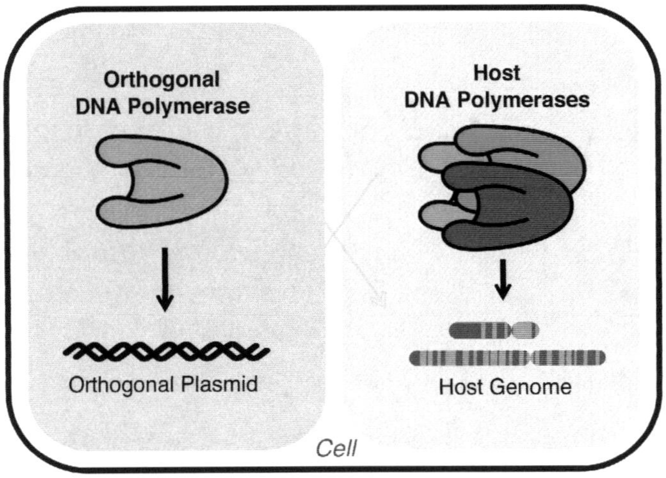

<u>The general reason</u>: Orthogonal systems are powerful platforms for synthetic biology because they allow us to freely engineer processes *in vivo* without influencing host processes in unintended ways.

Orthogonal Replication: Why?

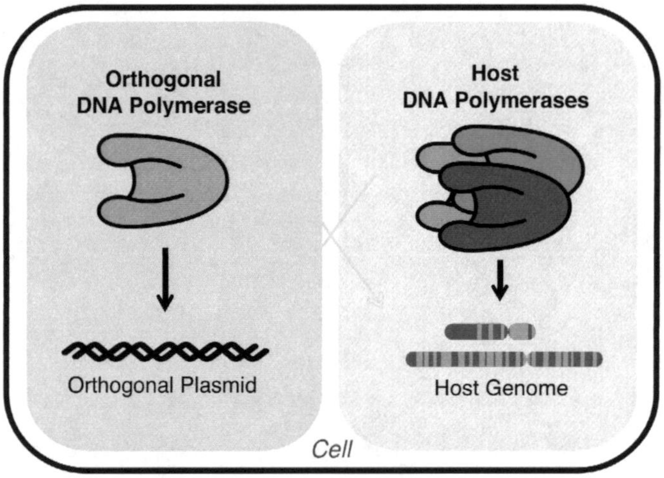

The general reason: Orthogonal systems are powerful platforms for synthetic biology because they allow us to freely engineer processes *in vivo* without influencing host processes in unintended ways.

Since DNA replication is responsible for transmission of biological information, orthogonal replication may have a special importance for synthetic biology.

Orthogonal Replication: Why?

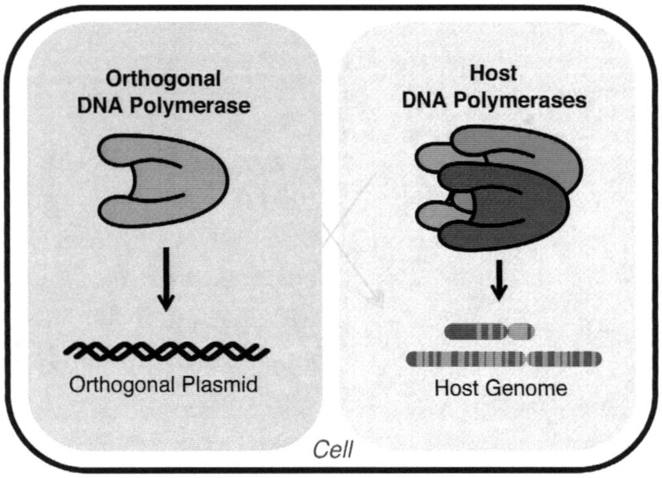

The specific reason for today's talk: Orthogonal replication can be a solution to the problem of continuous directed evolution *in vivo*.

Outline

1) **Introduction to continuous evolution**
2) **Implementation of orthogonal DNA replication for continuous evolution**
3) **Orthogonal replication: the larger picture**

Outline

1) **Introduction to continuous evolution**
2) Implementation of orthogonal DNA replication for continuous evolution
3) Orthogonal replication: the larger picture

The Difficulty of Engineering Biology

"In general, an engineer assumes that interacting components should be as simple as possible, that there are no "unnecessary" or unplanned interactions, that there is an explicit assignment of function or causal efficacy to each part of a working mechanism... Irrelevancy is avoided from the outset."

"By contrast, in evolutionary systems, where there is no design, the term "irrelevant" has no *a priori* meaning. It is possible for any change in a part to contribute to overall function, mutations can prompt compensation... there is no fixed assignment of exclusive responsibility for a given function, and, unlike the engineering case, interactions become increasingly complex."

Edelman & Gally, *PNAS*, 2001

The Difficulty of Engineering Biology

"In general, an engineer assumes that interacting components should be as simple as possible, that there are no "unnecessary" or unplanned interactions, that there is an explicit assignment of function or causal efficacy to each part of a working mechanism... Irrelevancy is avoided from the outset."

"By contrast, in evolutionary systems, where there is no design, the term "irrelevant" has no *a priori* meaning. It is possible for any change in a part to contribute to overall function, mutations can prompt compensation... there is no fixed assignment of exclusive responsibility for a given function, and, unlike the engineering case, interactions become increasingly complex."

Edelman & Gally, *PNAS*, 2001

So how do we engineer biology when surrounded by this complexity?

The Difficulty of Engineering Biology

"In general, an engineer assumes that interacting components should be as simple as possible, that there are no "unnecessary" or unplanned interactions, that there is an explicit assignment of function or causal efficacy to each part of a working mechanism... Irrelevancy is avoided from the outset."

"By contrast, in evolutionary systems, where there is no design, the term "irrelevant" has no *a priori* meaning. It is possible for any change in a part to contribute to overall function, mutations can prompt compensation... there is no fixed assignment of exclusive responsibility for a given function, and, unlike the engineering case, interactions become increasingly complex."

Edelman & Gally, *PNAS*, 2001

Use evolution: "...in biology, details matter a lot, and we don't understand the details. Evolution is a time-tested tool for engineering the details, and we can use it in the lab..."

Arnold & Meyerowitz, *Nature*, 2014

The Process: Evolution on Laboratory Timescales

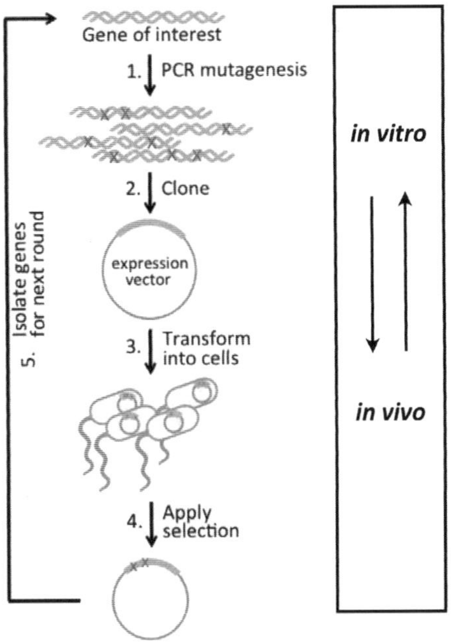

The Process: Evolution on Laboratory Timescales

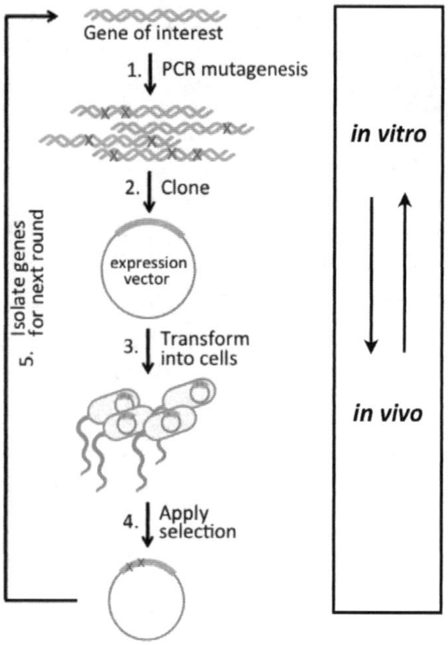

High mutation rate and **targeting**: We need to generate extensive diversity only in our genes of interest.

The Process: Evolution on Laboratory Timescales

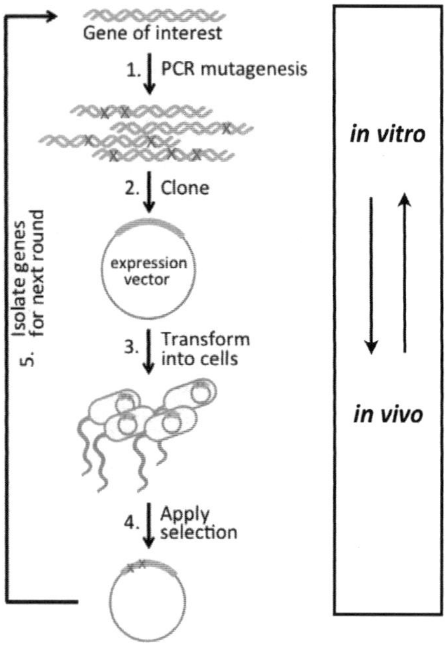

High mutation rate and **targeting**: We need to generate extensive diversity only in our genes of interest.

We need to express our genes, and they have to confer a selectable phenotype, which requires the cell.

The cellular environment is where we may employ our evolved gene in the end.

But…

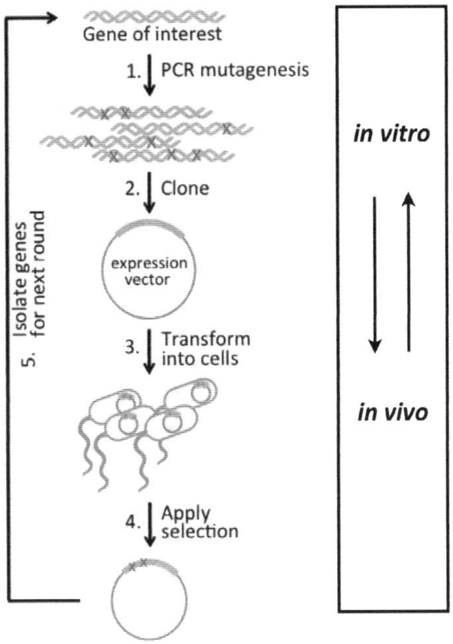

Relying on this stepwise back-and-forth between *in vitro* diversification and *in vivo* expression and selection, we lose two key properties of real evolution, which occurs **continuously** and is massively **parallel**.

But…

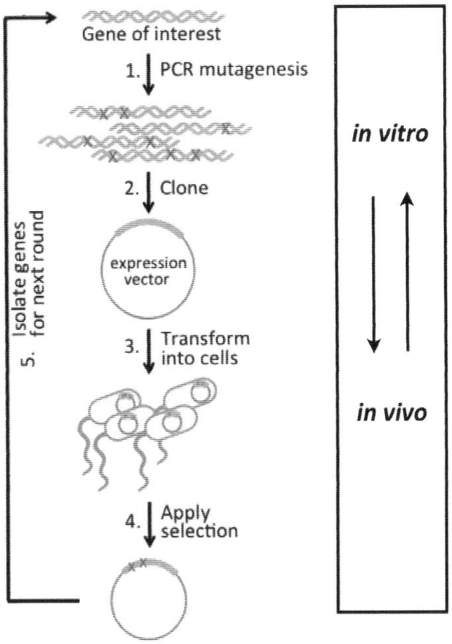

Relying on this stepwise back-and-forth between *in vitro* diversification and *in vivo* expression and selection, we lose two key properties of real evolution, which occurs **continuously** and is massively **parallel**.

Not continuous: only short exploration of adaptive pathways possible.

As we aim to evolve more complex or ambitious function, we will need to traverse long adaptive pathways.

Not parallelizable: limits statistical understanding; limits high-throughput experiments.

If we want to understand the scope of possible evolutionary trajectories (e.g. to map adaptive pathways in drug resistance), or want to evolve multiple activities (e.g. an series of biosensors), we need to evolve dozens or hundreds of lines in parallel.

Continuous Evolution

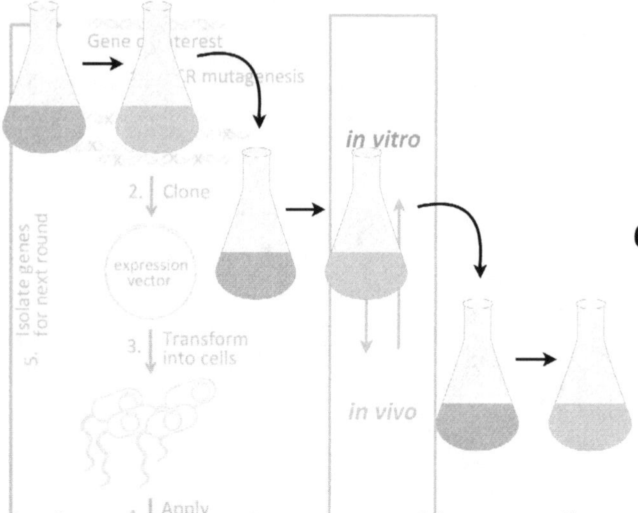

OR

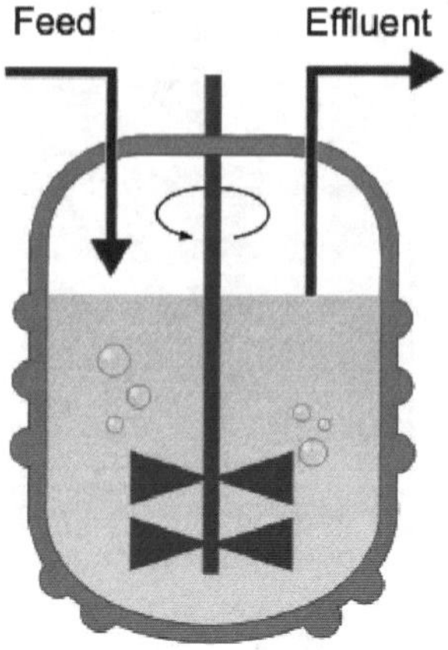

Replace this stepwise process with a more "natural" one, simply involving culturing cells over time.

Other groups: David Liu, Loeb, Maheshri, Cornish, Alper...

Continuous Evolution: Orthogonal Replication

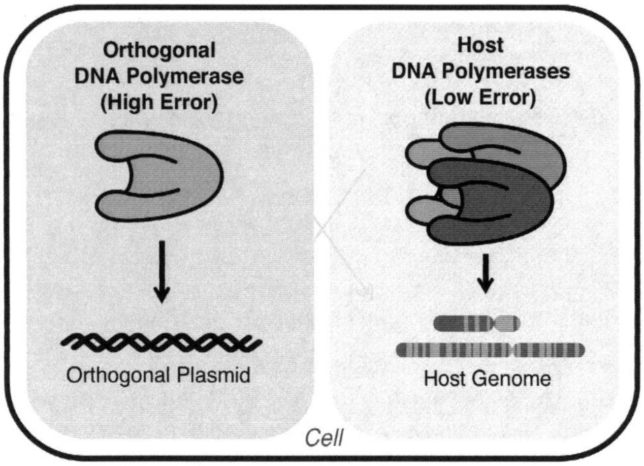

Ravikumar, Arrieta & Liu, *Nature Chem Biol*, 2014

Mutation rate: Can be high.

Targeting: Only genes of interest are rapidly mutated.

Continuous: Cells may undergo mutation, selection, and amplification as they propagate.

Parallelizable: Hundreds of evolving lines possible as the only labor involved is passaging cells.

Outline

1) **Introduction to continuous evolution**
2) Implementation of orthogonal DNA replication for continuous evolution
3) Orthogonal replication: the larger picture

Outline

1) Introduction to continuous evolution
2) **Implementation of orthogonal DNA replication for continuous evolution**
3) Orthogonal replication: the larger picture

Implementation

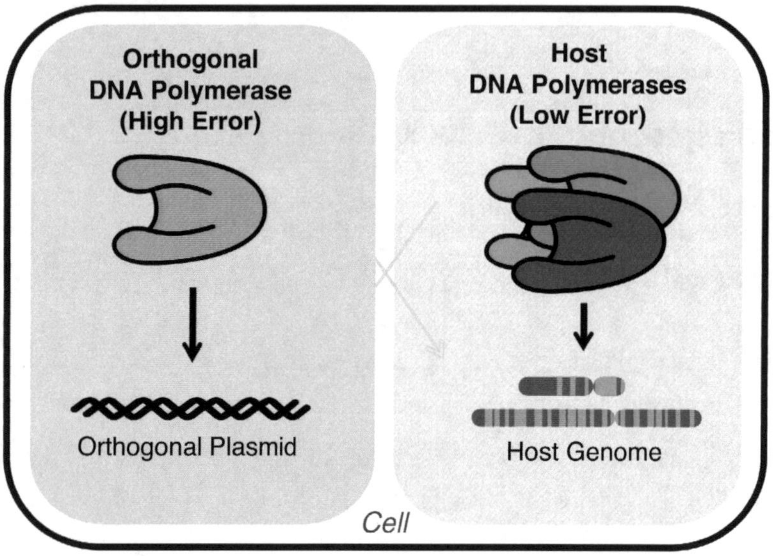

Implementation

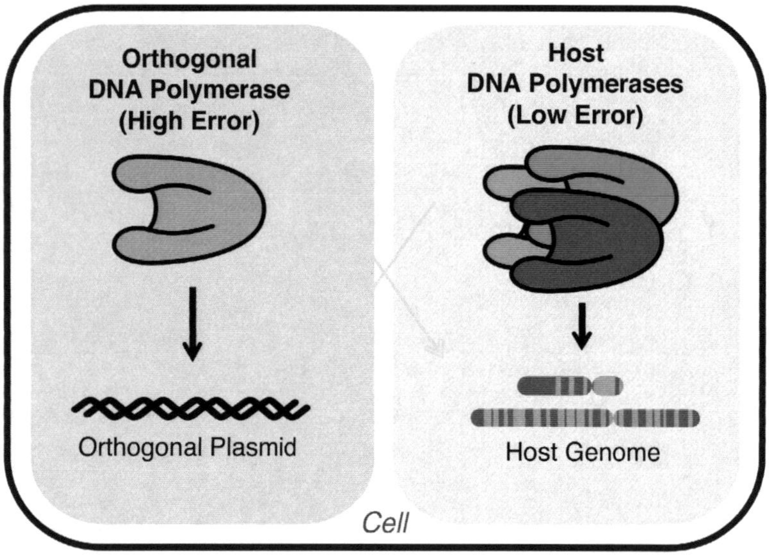

Borrow from nature

The pGKL1/pGKL2 Killer Plasmid System of *K. lactis*

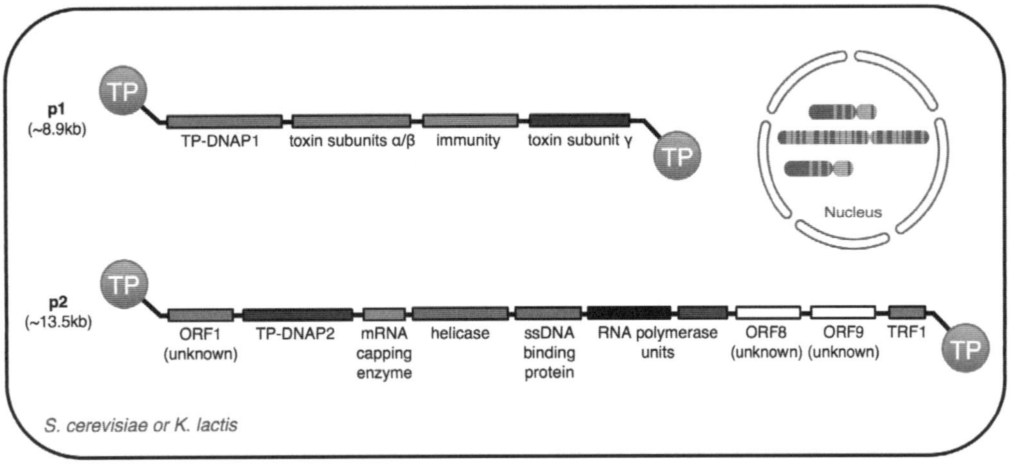

Gunge *et al.*, *J Bacteriol*, 1981

p1 encodes its own DNA polymerase (TP-DNAP1).

p2 also encodes its own DNA polymerase (TP-DNAP2) and a number of accessory protein necessary for replication of p1 and p2 and gene expression.

The pGKL1/pGKL2 Killer Plasmid System of *K. lactis*

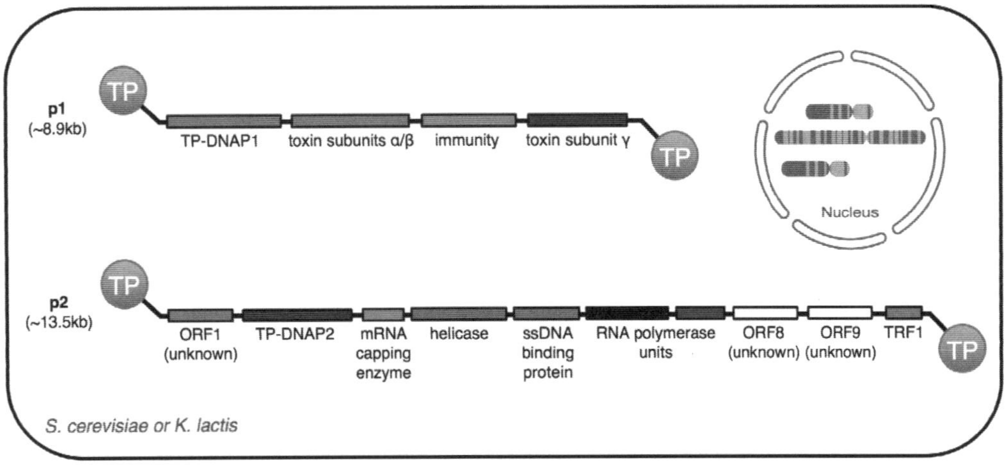

Gunge *et al.*, *J Bacteriol*, 1981

p1 encodes its own DNA polymerase (TP-DNAP1).

p2 also encodes its own DNA polymerase (TP-DNAP2) and a number of accessory protein necessary for replication of p1 and p2 and gene expression.

Three very attractive properties of this system...

The p1/p2 Selfish Replication System of *K. lactis*

1) An unusual replication mechanism: protein-primed replication.

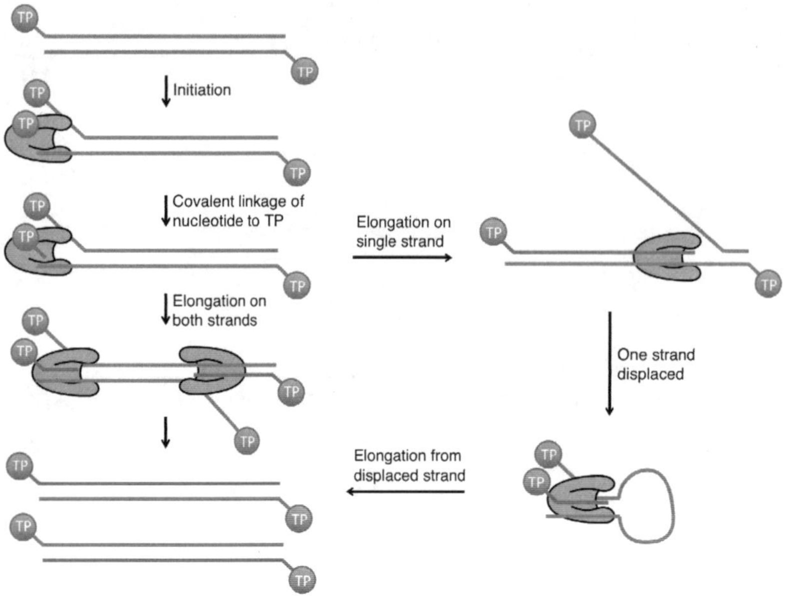

The p1/p2 Selfish Replication System of *K. lactis*

2) Replication occurs in the cytoplasm.

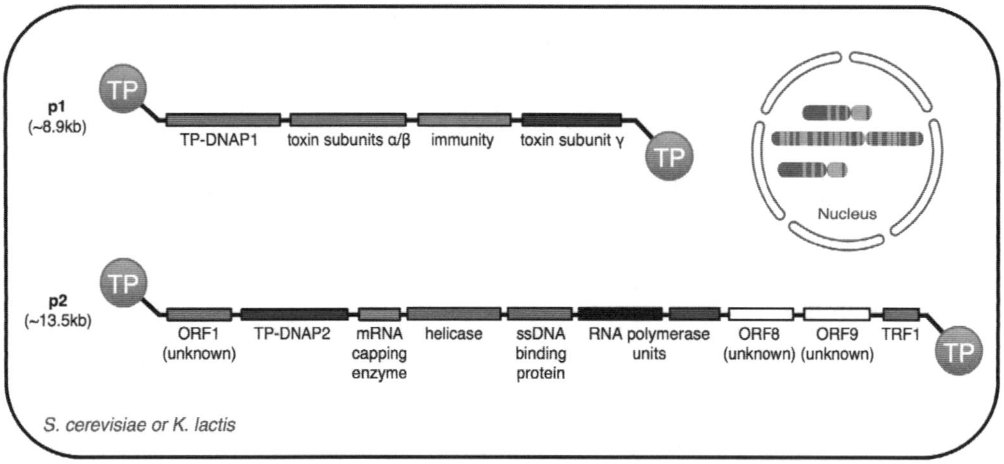

The p1/p2 Selfish Replication System of *K. lactis*

3) The system is autonomous (encoding everything needed) and minimal.

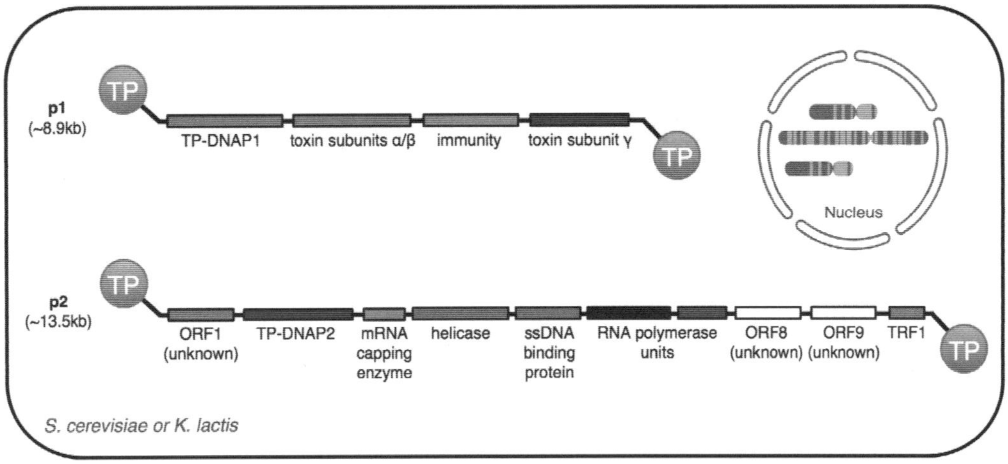

The p1/p2 Selfish Replication System of *K. lactis*

Unusual mechanism + cytoplasmic localization + autonomy implies p1/p2 is a parallel replication system independent of genomic replication

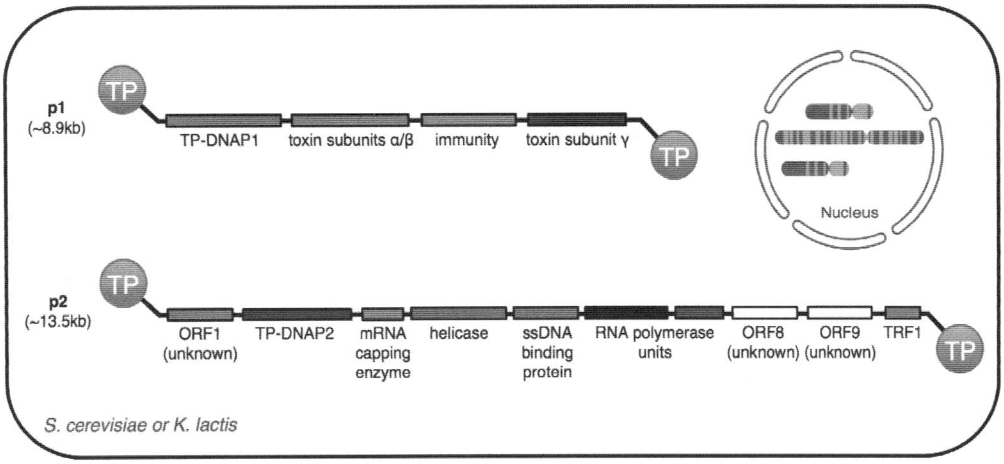

The p1/p2 Selfish Replication System of *K. lactis*

Unusual mechanism + cytoplasmic localization + autonomy implies p1/p2 is a parallel replication system independent of genomic replication

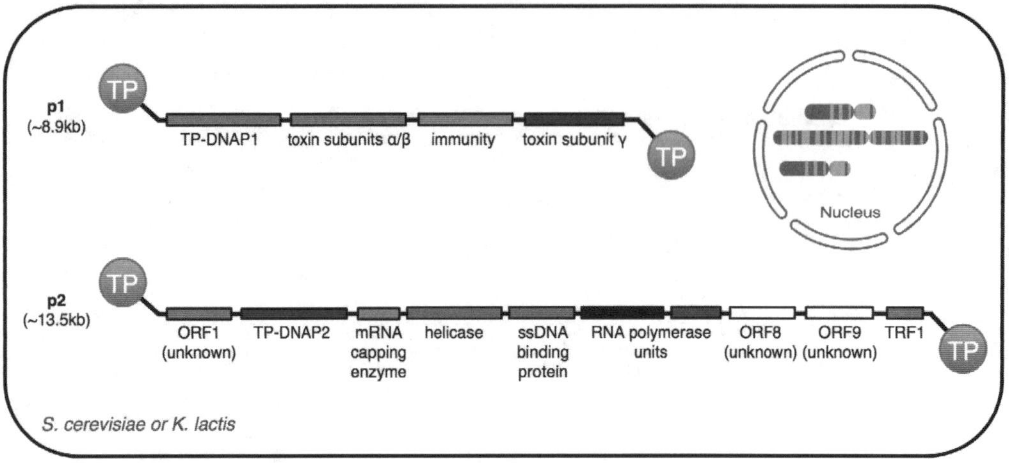

Implementation of orthogonal replication: develop the p1/p2 system

A Replication System Must Encode Genes of Interest

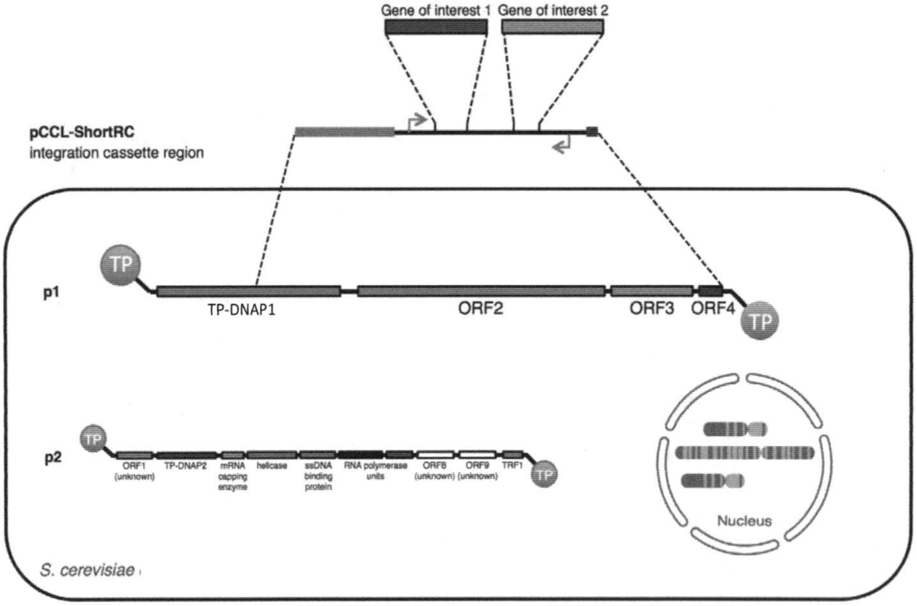

following Kamper *et al., Curr. Genet.*, 1991

DNA from Selected Integrations

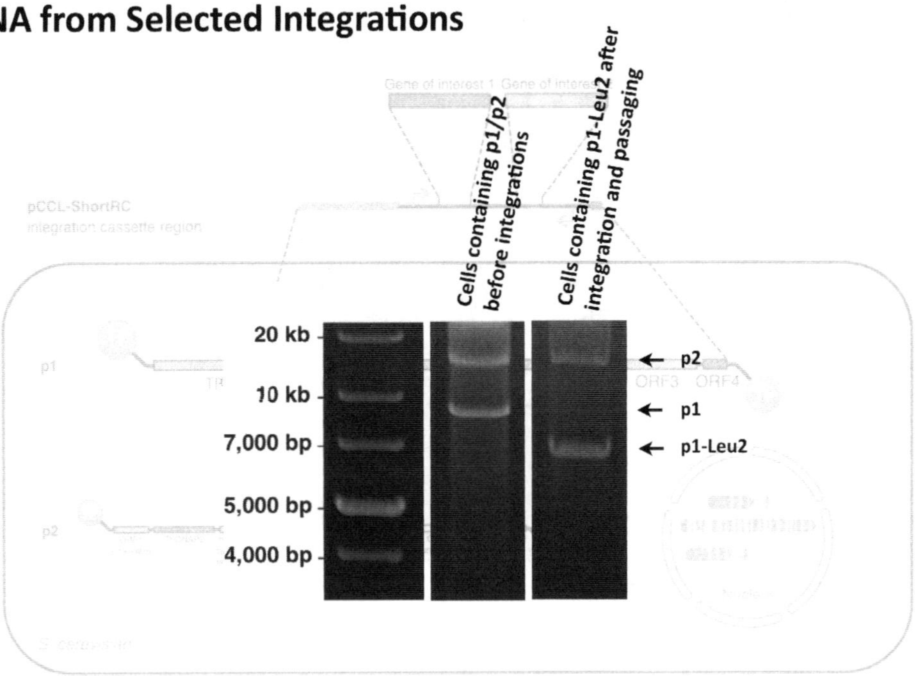

DNA from Selected Integrations

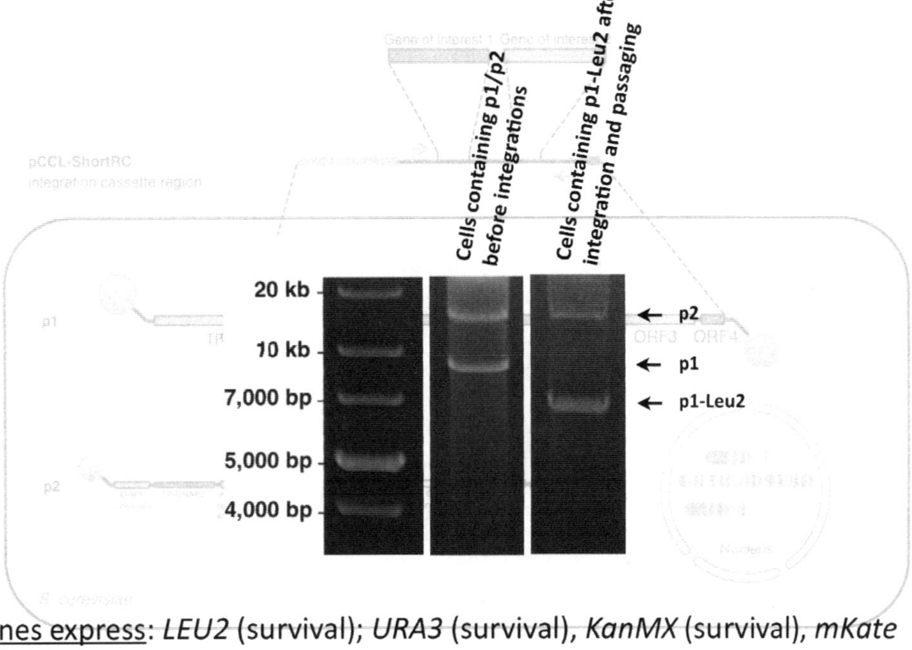

<u>Genes express</u>: *LEU2* (survival); *URA3* (survival), *KanMX* (survival), *mKate* (fluorescence), etc.

A Plan for Orthogonal Error-Prone Replication

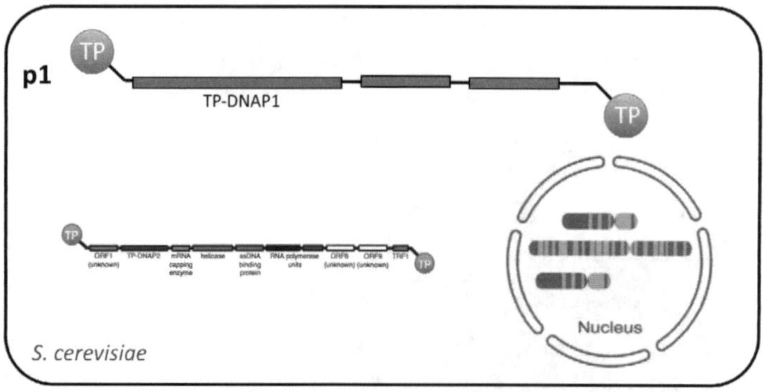

Can the mutation rate of p1 replication by TP-DNAP1 be high without an increase in the mutation rate of the host genome?

Can the mutation rate of the TP-DNAP1 be changed (made higher) without increasing the mutation rate of the host genome?

Determining the *in vivo* Error-Rate of p1

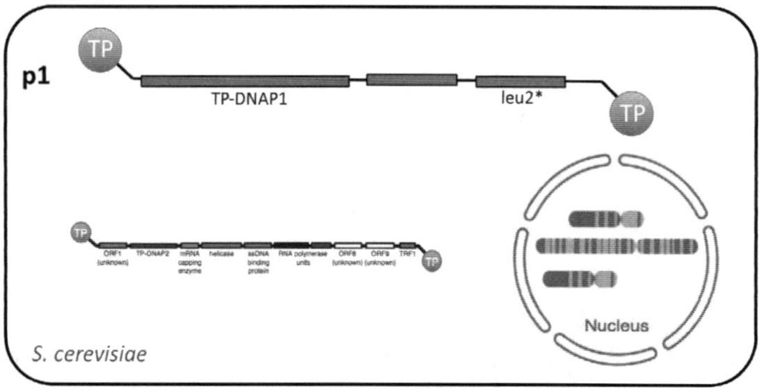

A reversion marker (*leu2**) integrated into p1

Fluctuation analysis using selection for *LEU2*

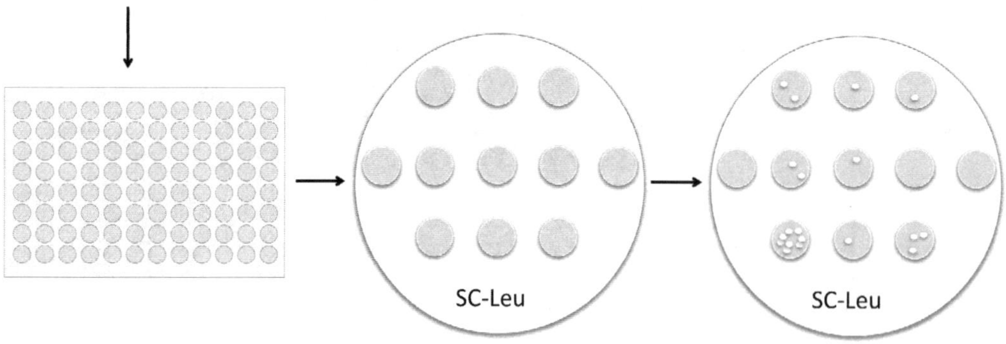

The *in vivo* Error-Rate of p1

46, 2, 6, 5, 4, 3, 5, 50, 3, 3, 50, 0, 60, 11, 150, 0, 4, 7, 0, 16, 6, 0, 3, 3, 9, 3, 0, 2, 0, 1, 1, 2, 0, 2, 0, 1, 4, 8, 0, 1, 0, 7, 2, 0, 4, 0, 5, 2, 2, 0, 1, 0, 0, 0, 0, 0, 0, 1, 3, 1, 5, 0, 0, 0, 0, 2, 1, 1, 2, 2, 13, 5, 0, 4, 1, 0, 0, 1, 10, 5, 4, 8, 10, 13, 2, 12, 4, 2, 5, 2, 56, 4, 12, 5, 4

The *in vivo* Error-Rate of p1

46, 2, 6, 5, 4, 3, 5, 50, 3, 3, 50, 0, 60, 11, 150, 0, 4, 7, 0, 16, 6, 0, 3, 3, 9, 3, 0, 2, 0, 1, 1, 2, 0, 2, 0, 1, 4, 8, 0, 1, 0, 7, 2, 0, 4, 0, 5, 2, 2, 0, 1, 0, 0, 0, 0, 0, 0, 1, 3, 1, 5, 0, 0, 0, 0, 2, 1, 1, 2, 2, 13, 5, 0, 4, 1, 0, 0, 1, 10, 5, 4, 8, 10, 13, 2, 12, 4, 2, 5, 2, 56, 4, 12, 5, 4

Description	# of replicates	Copy #	Per-base substitution rate (x 10^{-10})
p1 replication by wt TP-DNAP1	92	124	13.9 (10.9-17.1)

Number of mutation events (*m*) resulting in phenotype was determined from the number distribution of mutants using the Ma-Sandri-Sarkar maximal likelihood method.

m was converted to a per base mutation rate given 1) number of cells plated, 2) number of ways to revert, 3) copy number.

Copy number determined by qPCR.

Genomic Mutation Rate +/- p1 Replication

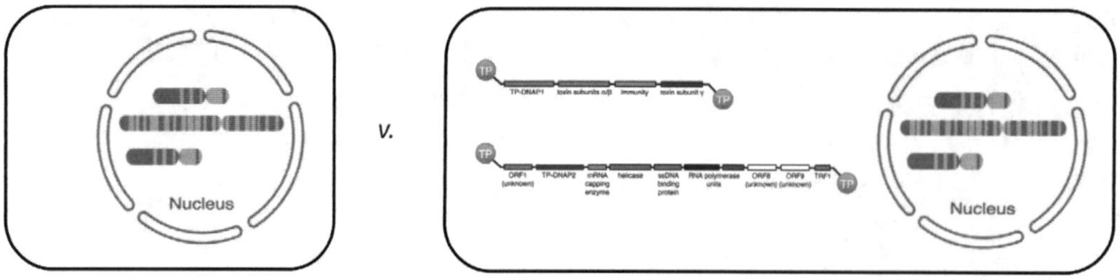

Description	# of replicates	Copy #	Per-base substitution rate (x 10^{-10})
Genomic rate in presence of p1 replication	183	1	1.29 (0.98-1.61)
Genomic rate in absence of p1 replication	92	1	1.59 (1.07-2.18)

Initial Mutation Rate Comparisons

Description	# of replicates	Copy #	Per-base substitution rate (x 10^{-10})
p1 replication by wt TP-DNAP1	92	124	13.9 (10.9-17.1)
Genomic rate in presence of p1 replication	183	1	1.29 (0.98-1.61)
Genomic rate in absence of p1 replication	92	1	1.59 (1.07-2.18)

p1's substitution mutation rate (1.39 x 10^{-9}) is 10-fold higher than genomic.

The presence of p1 replication doesn't increase the genomic mutation rate.

Initial Mutation Rate Comparisons

Description	# of replicates	Copy #	Per-base substitution rate (x 10^{-10})
p1 replication by wt TP-DNAP1	92	124	13.9 (10.9-17.1)
Genomic rate in presence of p1 replication	183	1	1.29 (0.98-1.61)
Genomic rate in absence of p1 replication	92	1	1.59 (1.07-2.18)

p1's substitution mutation rate (1.39 x 10^{-9}) is 10-fold higher than genomic.

The presence of p1 replication doesn't increase the genomic mutation rate.

Can the mutation rate of p1 replication by TP-DNAP1 be high without an increase in the mutation rate of the host genome? **Yes.** But 10-fold higher is meager.

Engineering Error-Prone DNA Polymerases

TP-DNAP1 is a family B DNA polymerase, most related to the Phi29 DNAP, and a lot is know about the fidelity of DNA polymerases

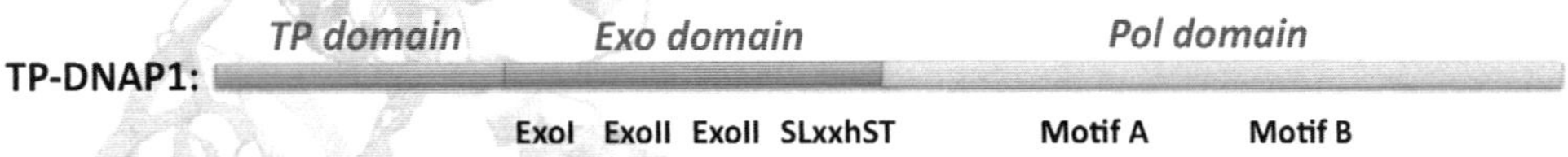

Exo I TVCFDIE**S**YFD (**S** to A, I, T) Exo II **N**IVD**F**EGR (**N** to A, D; **F** to A, S, Y) Exo II **N**GGG**Y**DFH (**N** to A, D; **Y** to A, S, Y) SLxxhST SFLLC**S**L (**S** to T)	Stabilization of primer terminus
Motif A DVKS**LY**P (**L** to ALL, **Y** to ALL) Motif B **YEKQN**N (**E** to A, **K** to A, **Q** to A, **N** to A)	Nucleotide recognition

Engineering Error-Prone DNA Polymerases

>100 mutants screened so far by fluctuation analysis

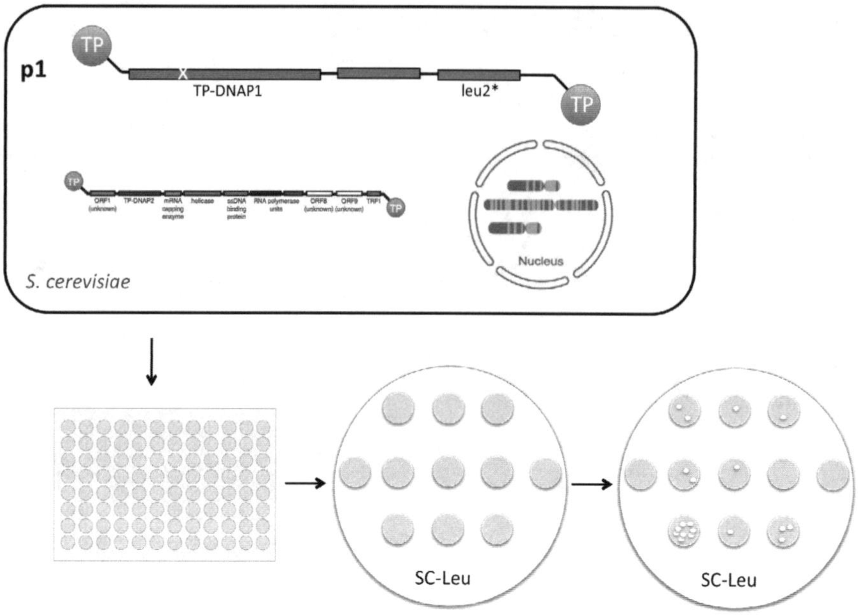

Expression of TP-DNAP1 *in trans*

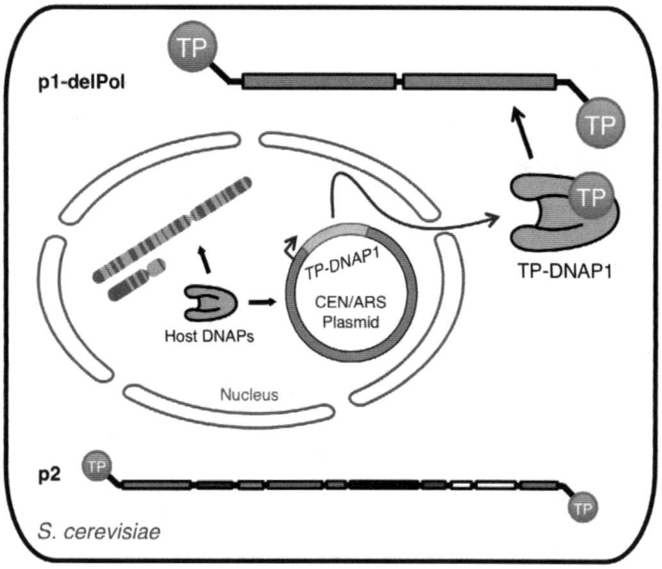

A codon-optimized, corrected TP-DNAP1 is expressed *in trans.*

TP-DNAP1 normally encoded on p1 is deleted.

Does TP-DNAP1 Replicate p1 *in trans*?

Does TP-DNAP1 Replicate p1 *in trans*?

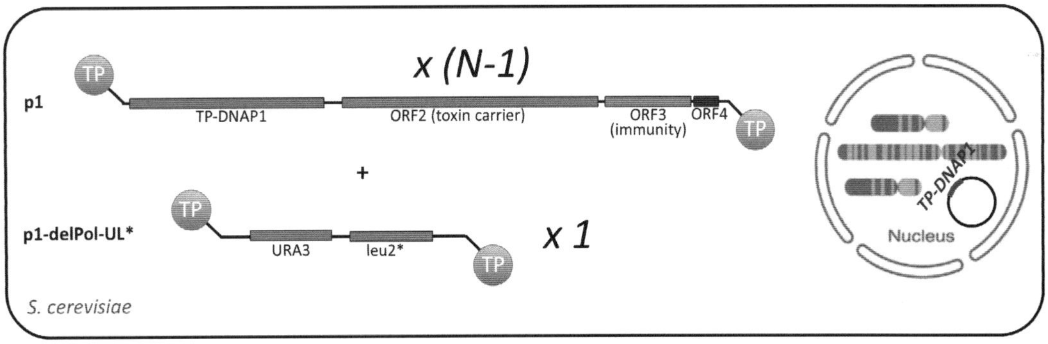

Selection for p1-delPol-UL* by growth in media without uracil.

Can copies of p1 disappear and can p1-delPol-UL* reach full copy number?

Complementation is Achieved

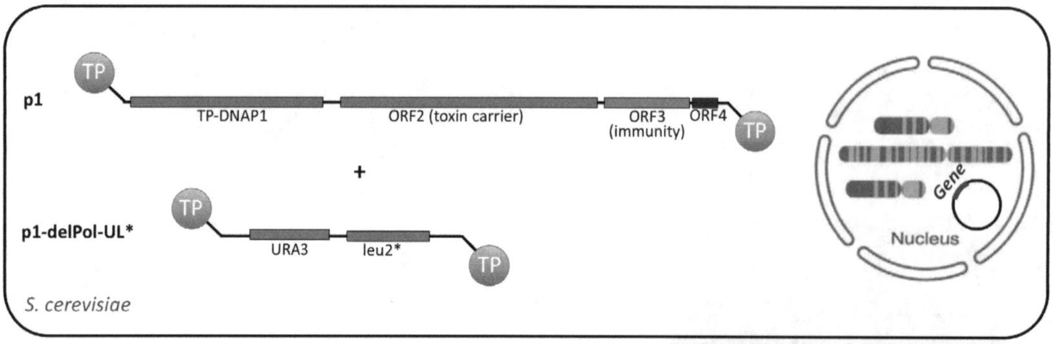

Selection for p1-delPol-UL* by growth in media without uracil.

Can copies of p1 disappear and can p1-delPol-UL* reach full copy number? **Yes**

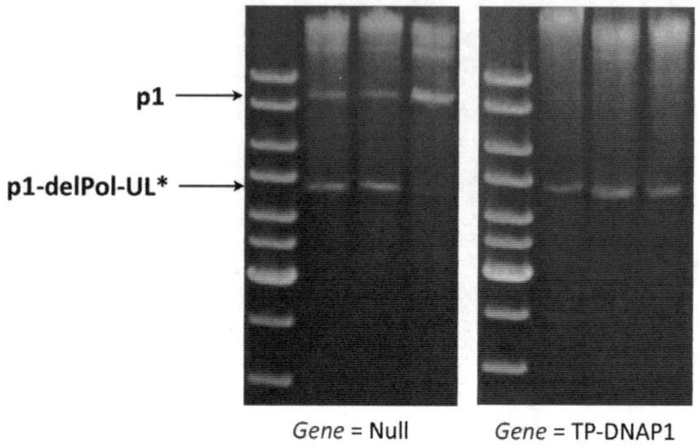

Towards Error-Prone Replication

Potential error-prone TP-DNAP1 mutants were expressed *in trans*, and error rates measured.

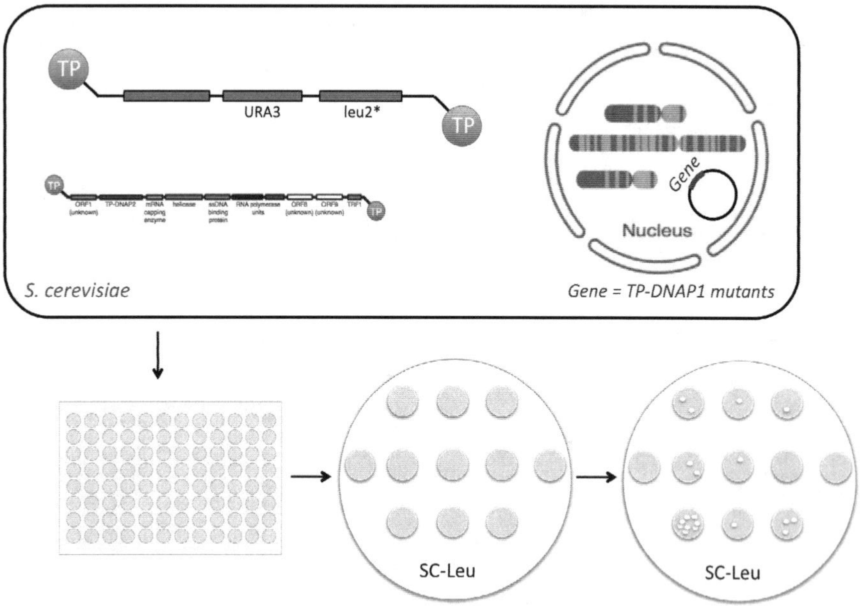

Error-Prone Replication

TP-DNAP1(Y427A) gave the highest mutation rate.

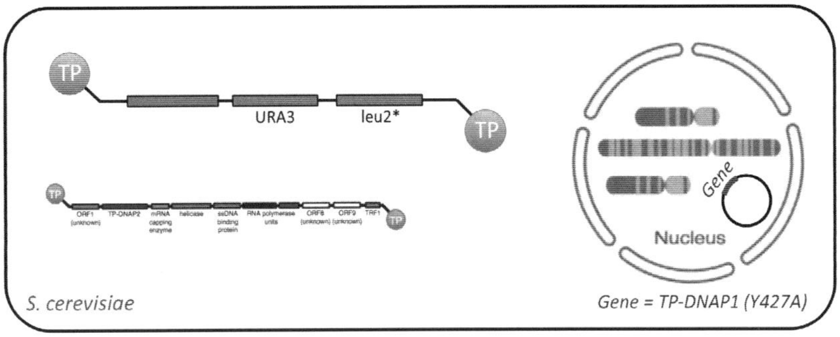

Description	# of replicates	Copy #	Per-base substitution rate (x 10^{-10})
p1 replication by wt TP-DNAP1	92	124	13.9 (10.9-17.1)
p1 replication by error-prone TP-DNAP1 *in trans*	68	52	350 (285-421)

Error-Prone Replication Stays Error-Prone

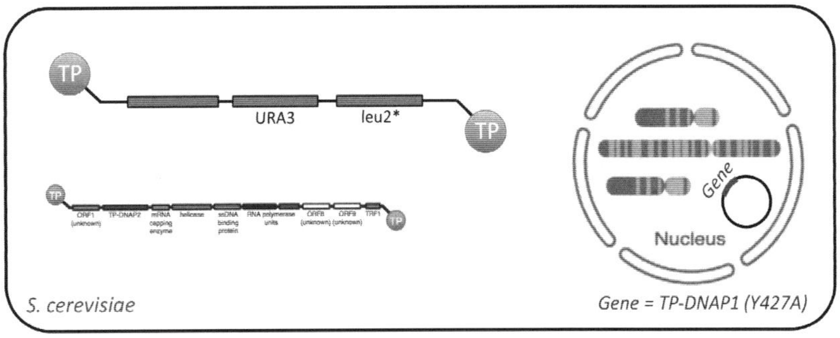

Description	# of replicates	Copy #	Per-base substitution rate (x 10^{-10})
p1 replication by error-prone TP-DNAP1 *in trans* (day 6, TDF=10^{12})	68	52	350 (285-421)
p1 replication by error-prone TP-DNAP1 *in trans* (day 8, TDF=10^{16})	60	52	316 (258-378)
p1 replication by error-prone TP-DNAP1 *in trans* (day 10, TDF=10^{20})	192	52	399 (351-447)
p1 replication by error-prone TP-DNAP1 *in trans* (day 18, TDF=10^{36})	94	52	371 (313-433)

Orthogonality

Genomic error rate in the absence and presence of the mutagenic TP-DNAP1

Orthogonality

Genomic error rate in the absence and presence of the mutagenic TP-DNAP1

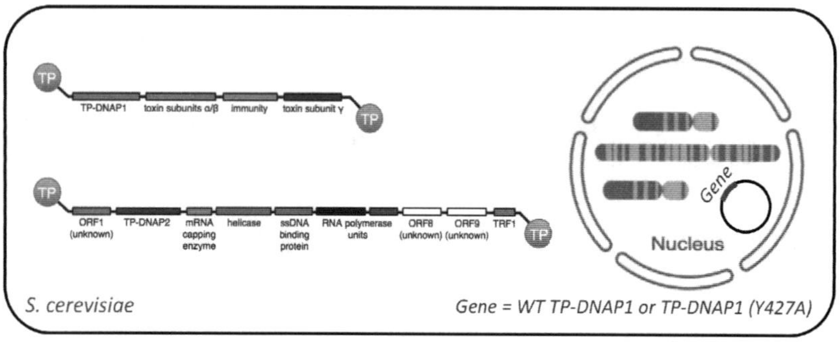

Description	# of replicates	Copy #	Per-base substitution rate (x 10^{-10})
Genomic rate in presence of p1 replication by wt TP-DNAP1 *in trans*	171	1	1.03 (0.77-1.32)
Genomic rate in presence of p1 replication by error-prone TP-DNAP1 *in trans*	177	1	0.78 (0.57-1.01)

Mutation Rate Comparisons

Description	# of replicates	Copy #	Per-base substitution rate (x 10^{-10})
p1 replication by mutant TP-DNAP1 *in trans*	68	52	350 (285-421)
Genomic rate in presence of p1 replication by wt TP-DNAP1 *in trans*	171	1	1.03 (0.77-1.32)
Genomic rate in presence of p1 replication by error-prone TP-DNAP1 *in trans*	177	1	0.78 (0.57-1.01)

Using an engineered error-prone TP-DNAP, p1's mutation rate (3.5×10^{-8}) is 400-fold higher than genomic.

The presence of p1 replication by an error-prone TP-DNAP1 doesn't increase the genomic mutation rate.

Can the mutation rate of p1 replication by TP-DNAP1 be made higher without an increase in the mutation rate of the host genome? **Yes.**

Orthogonal Error-Prone Replication Version 1.0

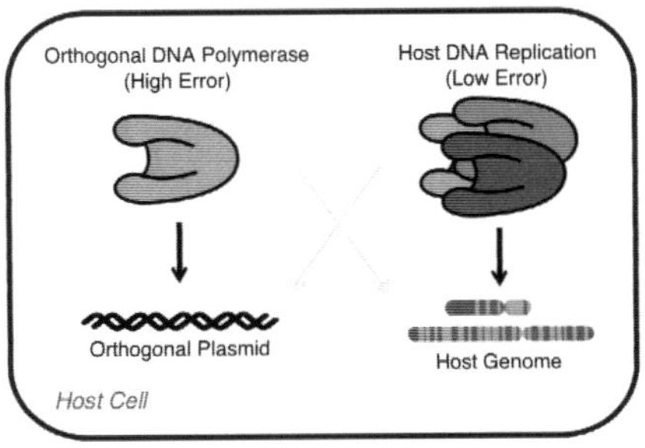

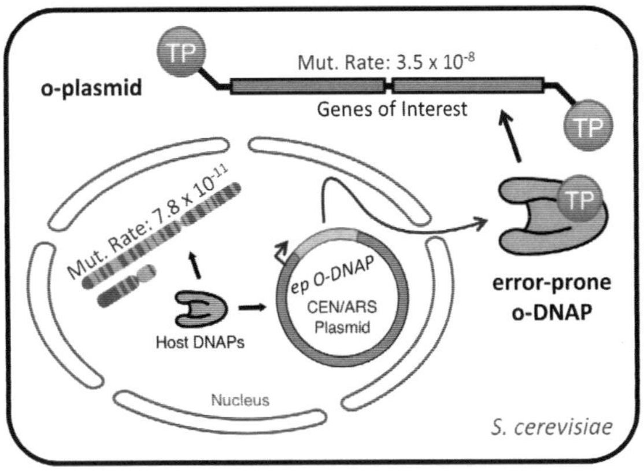

Ravikumar, Arrieta & Liu, *Nature Chem Biol*, 2014

Continuous evolution *in vivo*: at a copy number of 52, this means that single point mutants of any gene will be generated ~23,000-fold more quickly if encoded on p1 instead of the yeast genome.

Targets

We have encoded several targets on the ortho-plasmid and are setting up selections

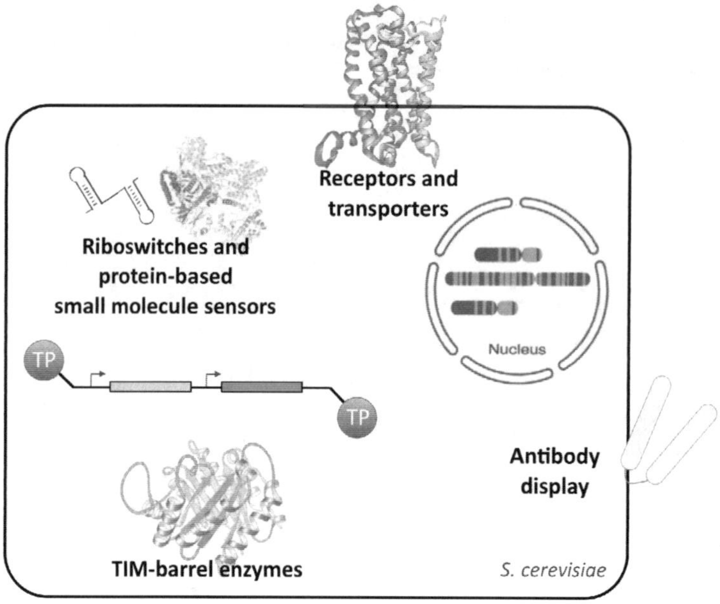

Ongoing Work

New and higher error rates

Large library of DNAPs being screened for higher error-rates against numerous genetic backgrounds.

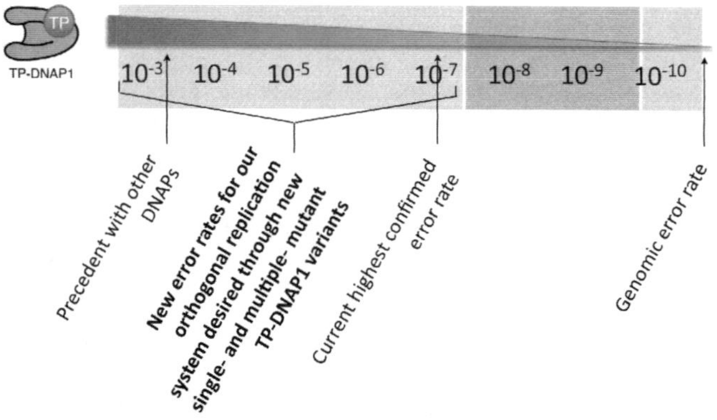

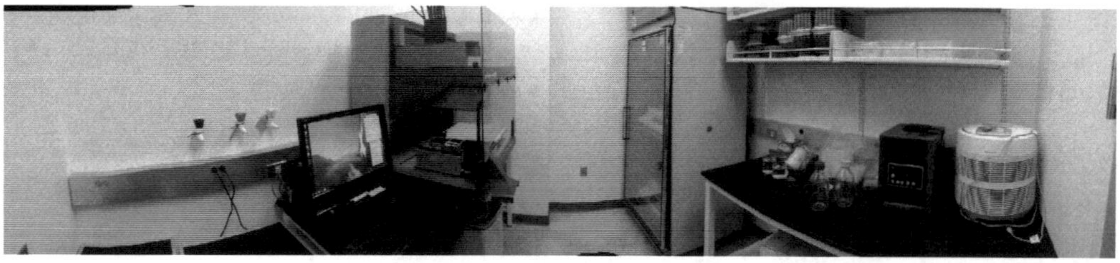

Ongoing Work

Copy number control

Significance: Under current high copy numbers, a mutation must be dominant positive to be selectable, and counterselections are difficult. Lowering copy number is desirable.

Approach: Express a dummy orthogonal DNA polymerase (D641A mutant) that competes for initiation but doesn't polymerize DNA.

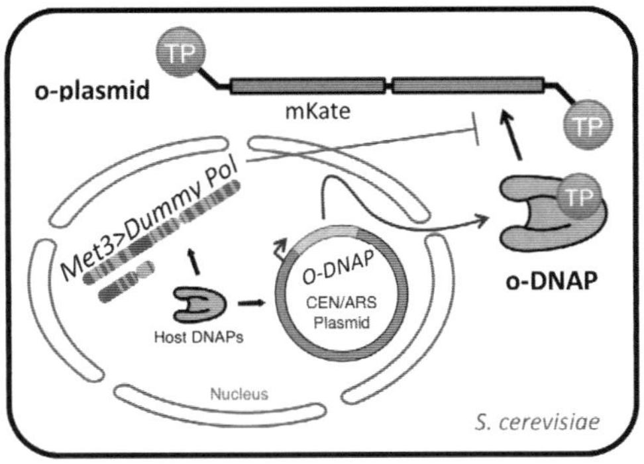

The Met3 promoter is repressed by methionine

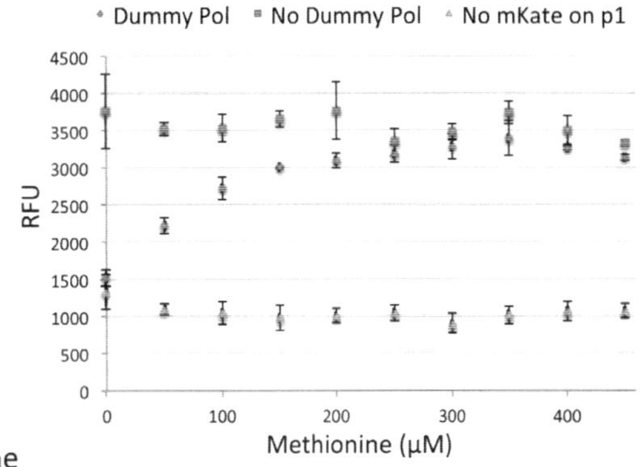

Ongoing Work

Sexual recombination

Significance: Rapid *in vivo* mutation of the o-plasmid supports asexual selection, but sexual selection is a critical process for evolutionary innovation that combines beneficial mutations found in separate lineages.

Approach: Transfer p1/p2 into mating strains and mate separately evolving lineages as desired.

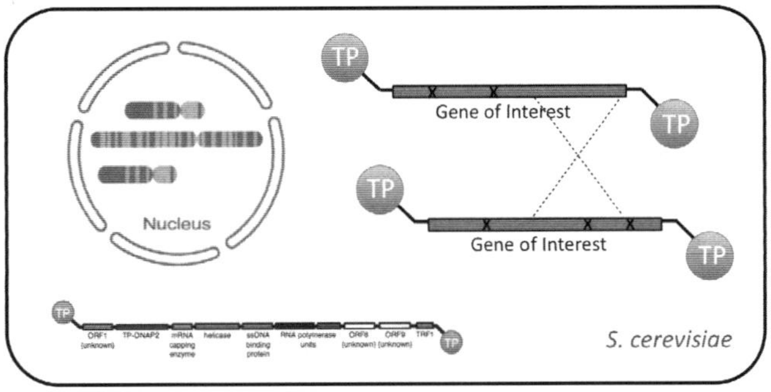

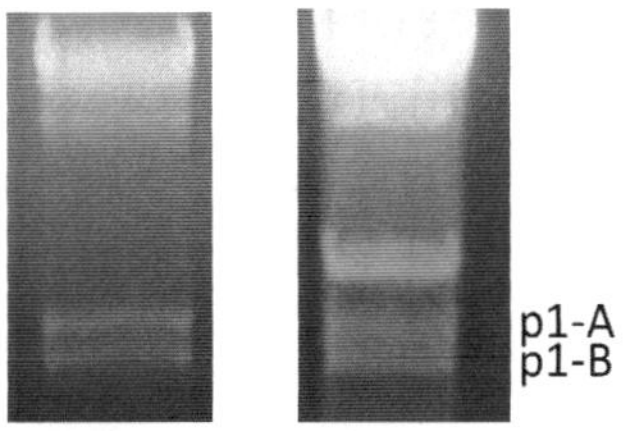

Before recombination

After recombination

Measured recombination rates (after 24 hours growth): 30bp overlap — 10^{-7} events/cell; 100bp overlap — 10^{-5} events/cell; 500bp overlap — over 10^{-3} events/cell

Outline

1) Introduction to continuous evolution

2) **Implementation of orthogonal DNA replication for continuous evolution**

3) Orthogonal replication: the larger picture

Outline

1) Introduction to continuous evolution

2) Implementation of orthogonal DNA replication for continuous evolution

3) **Orthogonal replication: the larger picture**

Orthogonal Replication

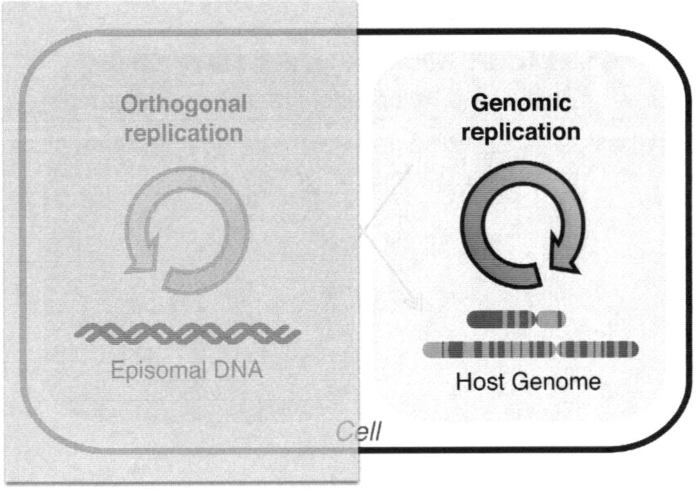

The genome imposes strict limits on our ability to engineer how DNA information storage and transmission occurs. There is no room for creativity.

Orthogonal Replication

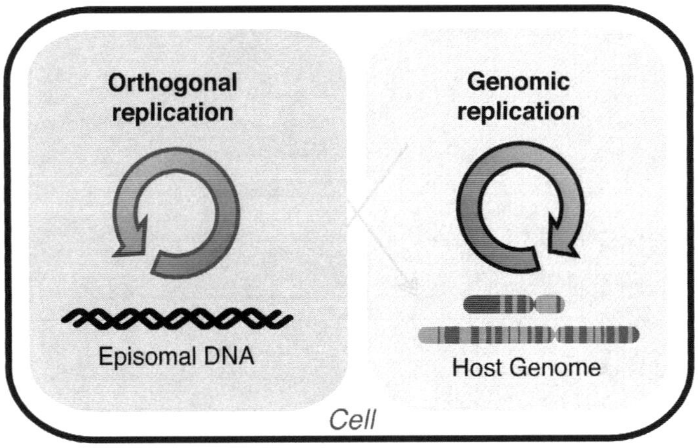

The genome imposes strict limits on our ability to engineer how DNA information storage and transmission occurs. There is no room for creativity.

Our orthogonal system need not be used for the essential task of accurately copying existing information, so we can develop new ways to record and transmit DNA information.

Questions

How much can we accelerate mutation *in vivo*?

There is a roughly inverse relationship between genome size and maximum tolerable mutation rate. A kilobase gene should be able to sustain much higher mutation rates than a megabase-gigabase genome. Orthogonal replication lets us evolve the kilobase gene at the much higher rate it can tolerate.

How exceptional are the chemistries of life?

Orthogonal replication is a platform for building XNA replication (using unnatural base pairs and nucleotide backbones) *in vivo*.

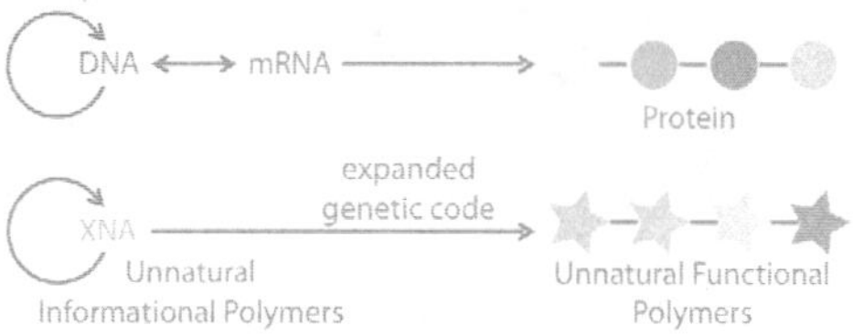

Can we record non-genetic environmental information as heritable genetic information?

Orthogonal polymerases that respond to arbitrary cellular stimuli and record events as sequenceable genetic changes in the orthogonal plasmid.

Questions

How much can we accelerate mutation *in vivo*?

There is a roughly inverse relationship between genome size and maximum tolerable mutation rate. A kilobase gene should be able to sustain much higher mutation rates than a megabase-gigabase genome. Orthogonal replication lets us evolve the kilobase gene at the much higher rate it can tolerate.

How exceptional are the chemistries of life?

Orthogonal replication is a platform for building XNA replication (using unnatural base pairs and nucleotide backbones) *in vivo*.

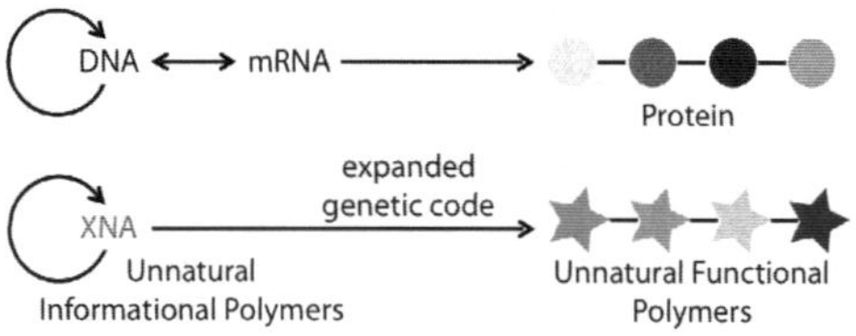

Can we record non-genetic environmental information as heritable genetic information?

Orthogonal polymerases that respond to arbitrary cellular stimuli and record events as sequenceable genetic changes in the orthogonal plasmid.

Questions

How much can we accelerate mutation *in vivo*?

There is a roughly inverse relationship between genome size and maximum tolerable mutation rate. A kilobase gene should be able to sustain much higher mutation rates than a megabase-gigabase genome. Orthogonal replication lets us evolve the kilobase gene at the much higher rate it can tolerate.

How exceptional are the chemistries of life?

Orthogonal replication is a platform for building XNA replication (using unnatural base pairs and nucleotide backbones) *in vivo*.

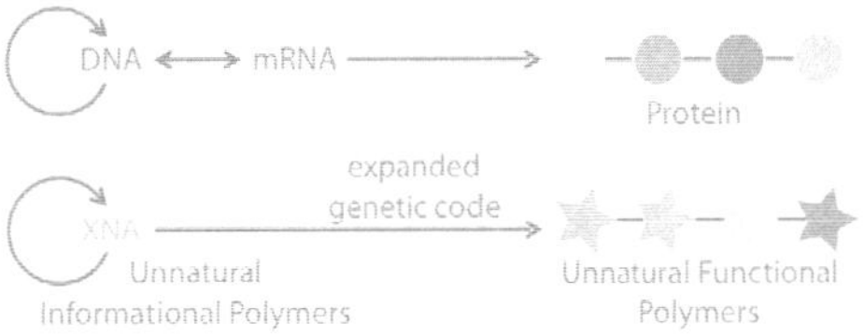

Can we record non-genetic environmental information as heritable genetic information?

Orthogonal polymerases that respond to arbitrary cellular stimuli and record events as sequenceable genetic changes in the orthogonal plasmid.

Acknowledgements

Arjun Ravikumar

Adrian Arrieta

Alex Javanpour

Ziwei Zhong

Reverse engineering enzymes with high-throughput sequence-function mapping

Philip A. Romero
University of California, San Francisco

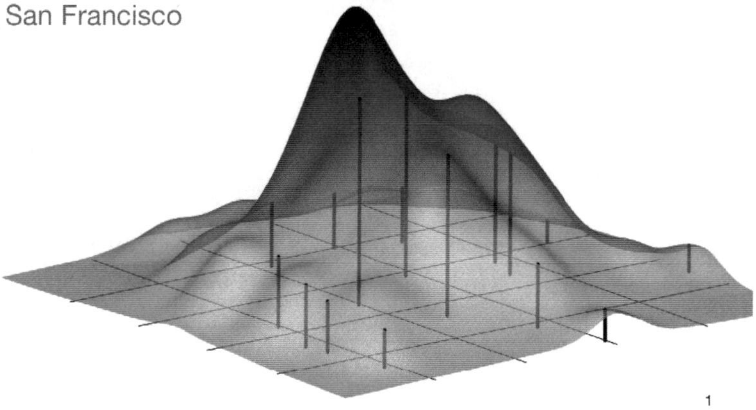

1

Enzymatic rate acceleration

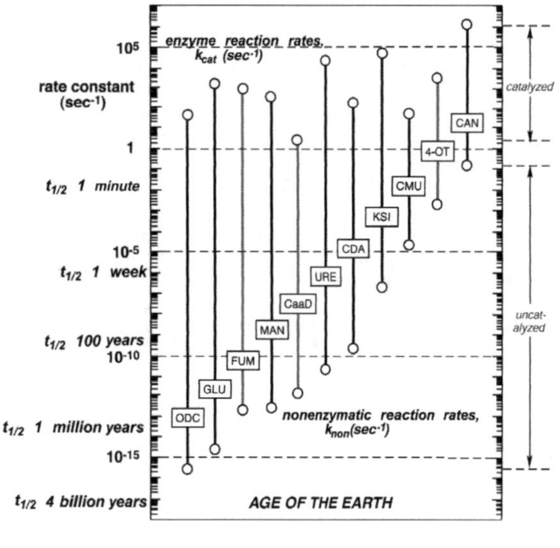

Horvat, *PNAS* 2005

- Mechanisms are complex
  - multistep
  - multiple catalytic mechanisms
  - rely on numerous molecular interactions
- Limits our understanding of the relationship between enzyme sequence and function

2

Deep mutational scanning

- Combining high-throughput functional characterization with deep sequencing
  - phage display
  - growth selections
- Large, unbiased sequence-function data sets
  - study structure of fitness landscape
  - discover new functional sites
  - improve energy functions
  - identify beneficial mutations
- <u>Limitation:</u> most proteins can't be coupled to binding or cellular growth

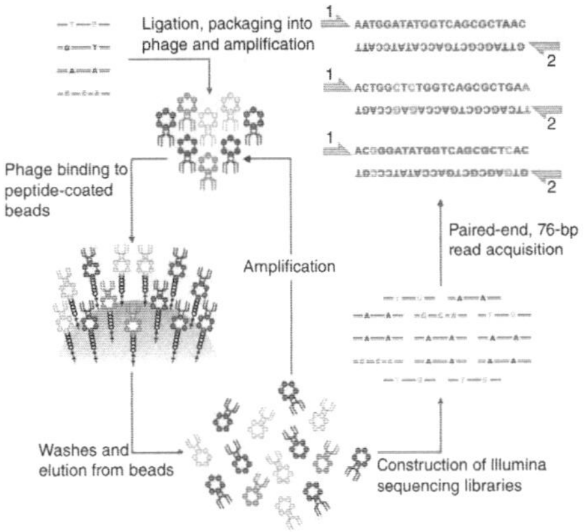

Fowler et al. *Nat Meth* 2010, 2014

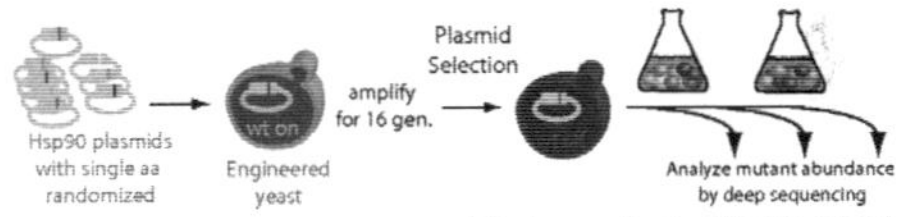

Hietpas et al. *PNAS* 2011

Microemulsions as vessels for biological reactions

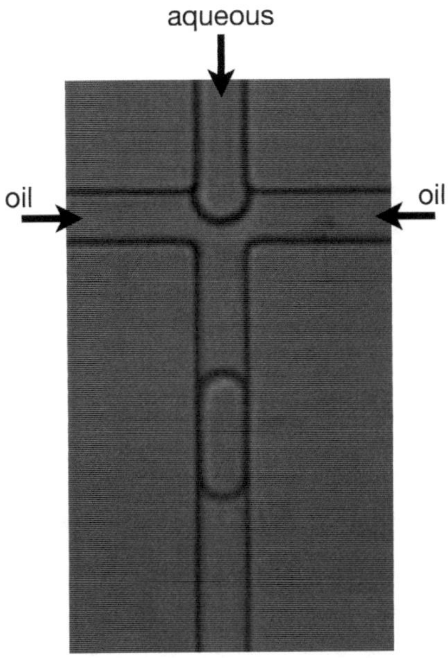

Compartmentalized biological reactions

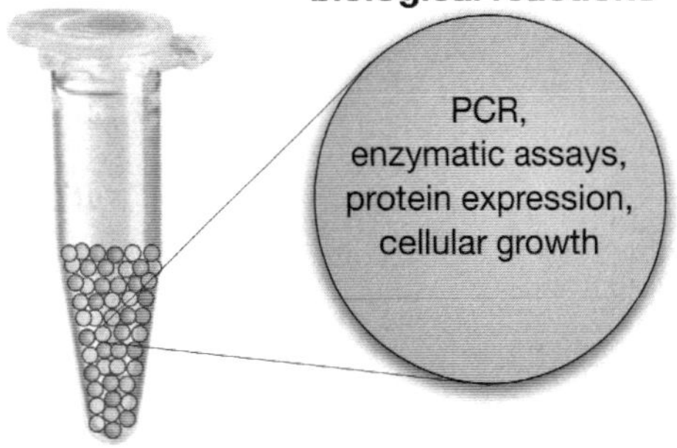

- Small volumes: 1 pL - 1 nL
- Monodisperse: ~1% size variation
- Rapid: drop generation and processing at kHz rates
- Microfluidic automation

4

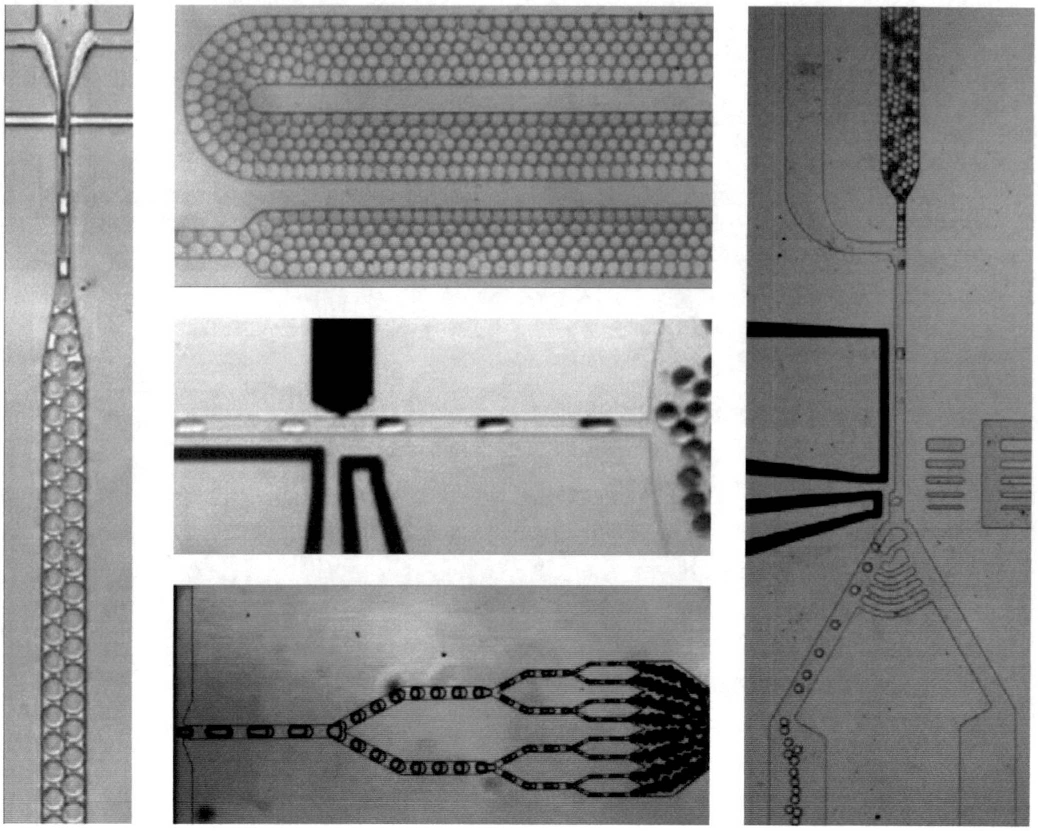

Ultrahigh-throughput screening of enzyme libraries

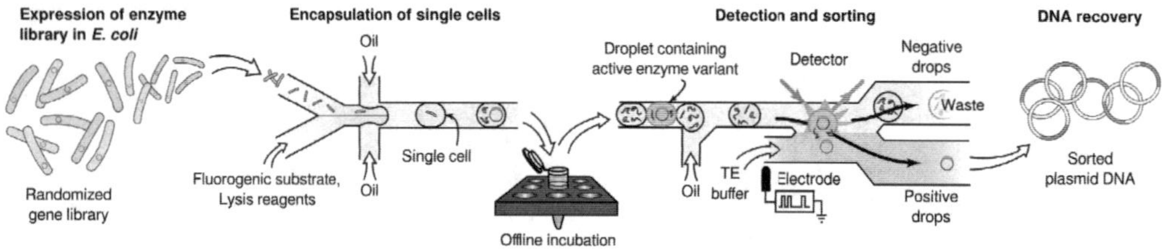

- Screening one million enzyme variants
  - a few hours
  - less than 100 μL reagents
- Platform technology
  - directed evolution
  - functional metagenomic screening
  - comprehensive mapping of sequence-function relationships

6

High-throughput mapping of enzyme sequence-function relationships

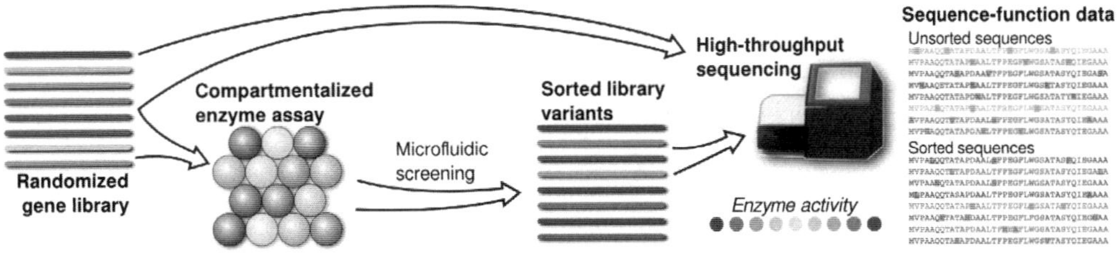

Combine microfluidic screening and next-gen DNA sequencing technologies to generate massive <u>enzyme</u> sequence-function data sets

- dissect the molecular basis of enzyme function
- identify beneficial mutations for enzyme engineering

7

Emulsion-based beta-glucosidase assay

- Glycoside hydrolase family 1 (GH1)
  - hydrolyze β-glycosidic bonds between two or more carbohydrates
  - Beta-glucosidase activity: important for biomass deconstruction
  - Bgl3 from *Streptomyces sp.* $k_{cat}/k_{uncat} \approx$ 1e16

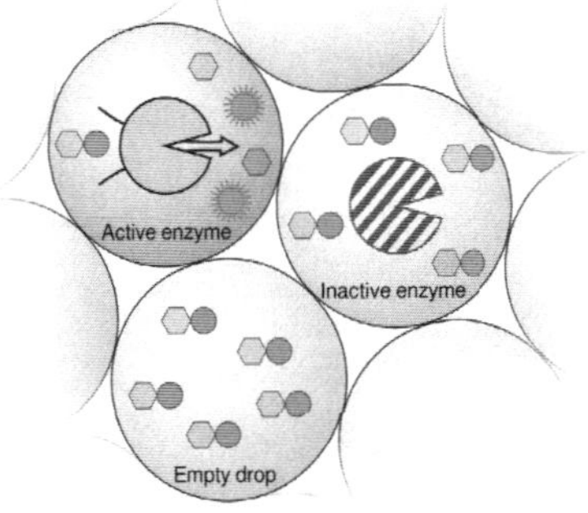

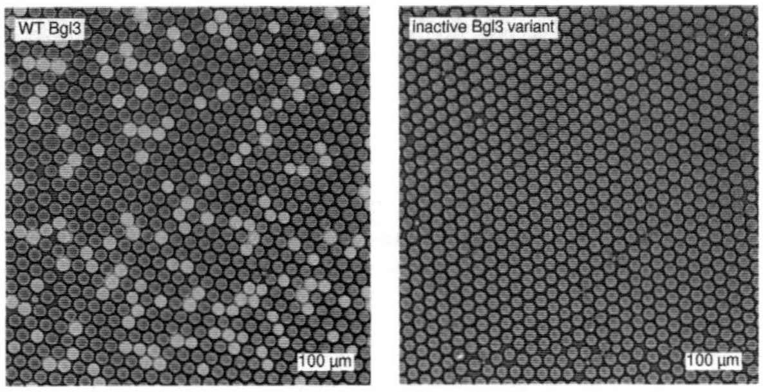

Poisson encapsulation (λ = 0.1)

8

Microfluidic screening of beta-glucosidase library

<u>Error-prone PCR library</u>
6 million mutants
3.9 AA subs./gene
35% active

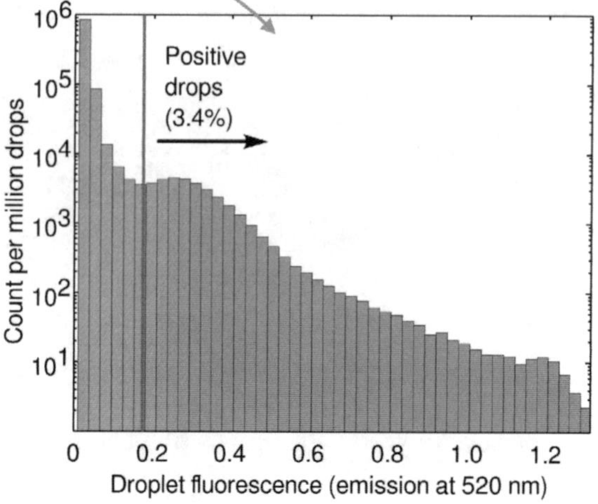

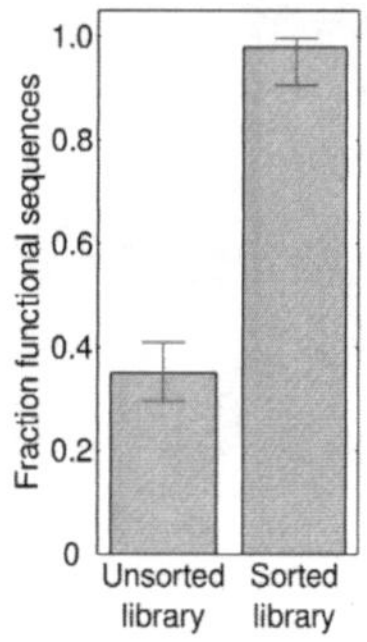

sort 10 million active variants

Analyzing > one million sequences!

Frequency of amino acid substitutions

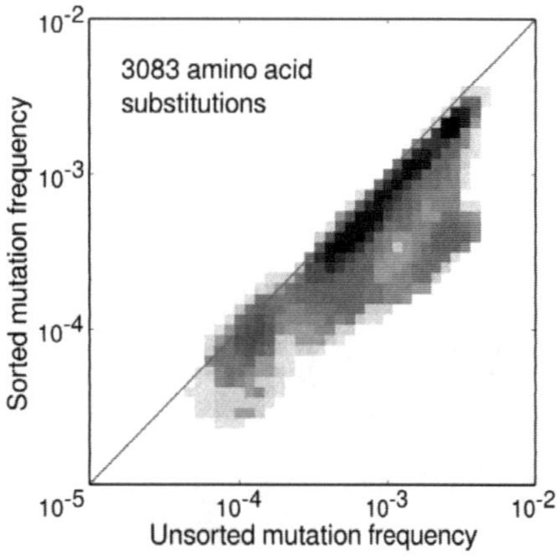

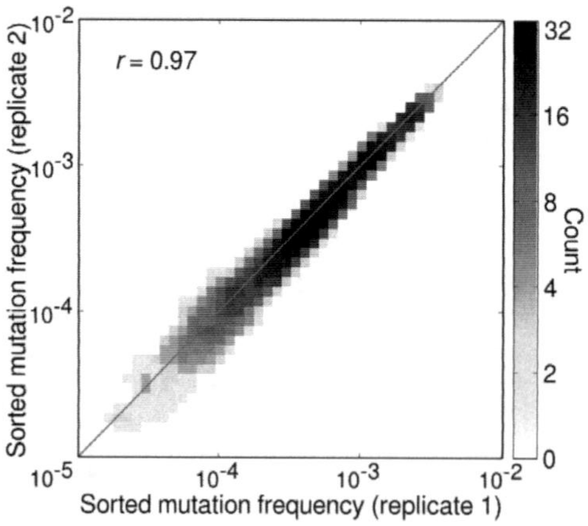

Site-specific mutational tolerance

- Relative entropy

$$RE = \sum_a p_{a,sort} \log \left(\frac{p_{a,sort}}{p_{a,pre}} \right)$$

  - quantifies how much amino acid distribution changes upon selection
  - high *RE*: important residue

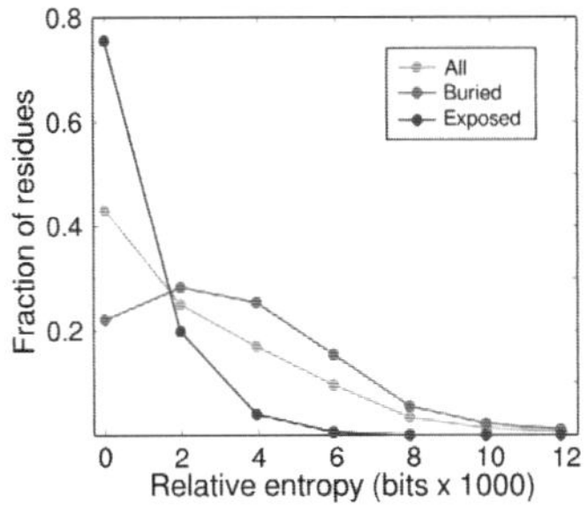

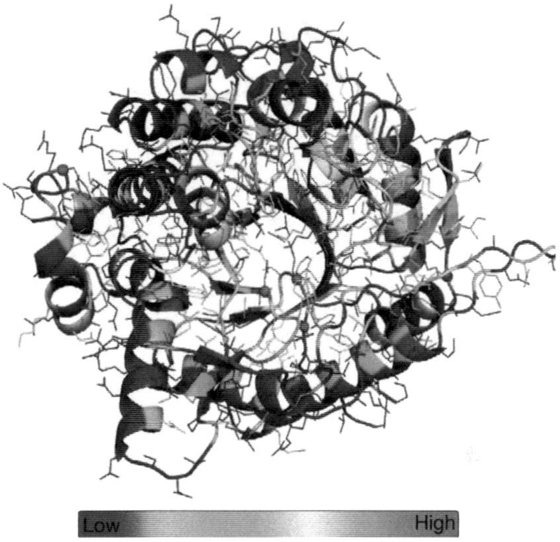

11

Analysis reveals new sites important for function

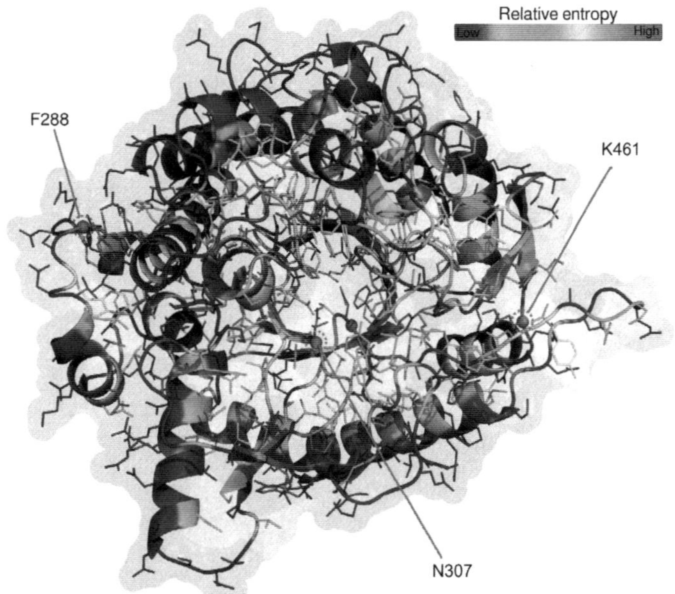

12

Position K461

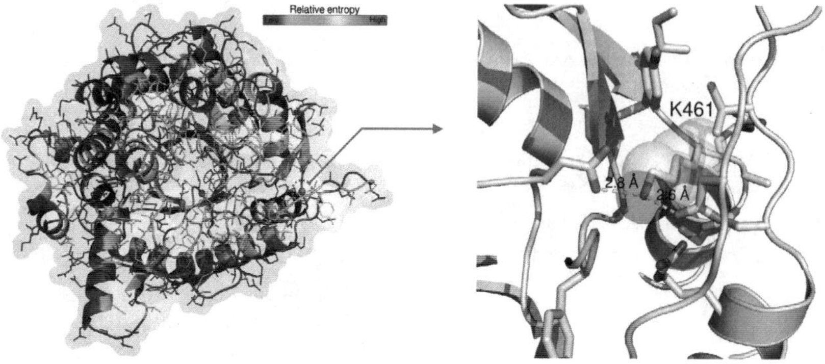

mutagenesis at 461

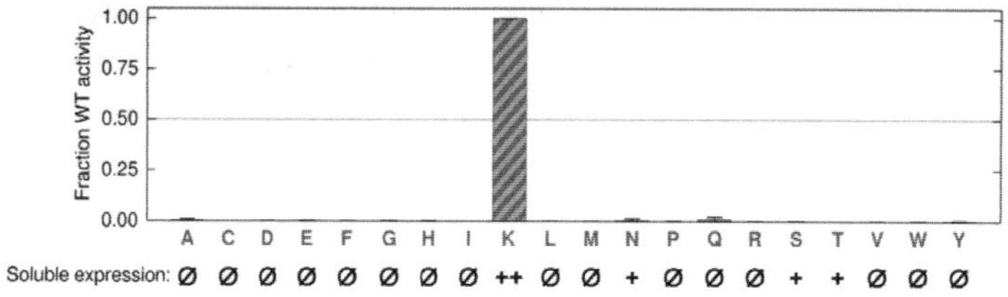

13

Position N307

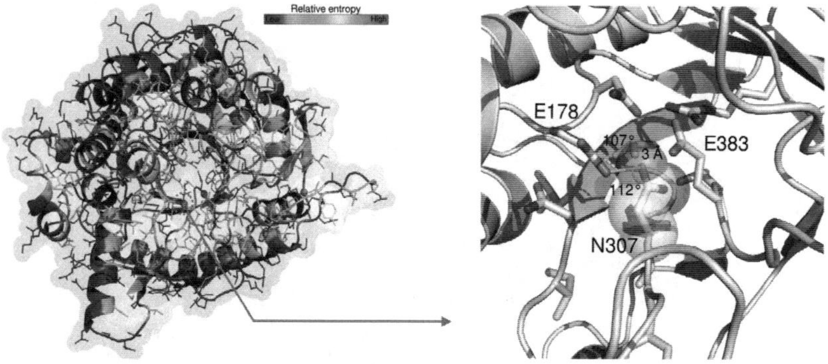

mutagenesis at 307

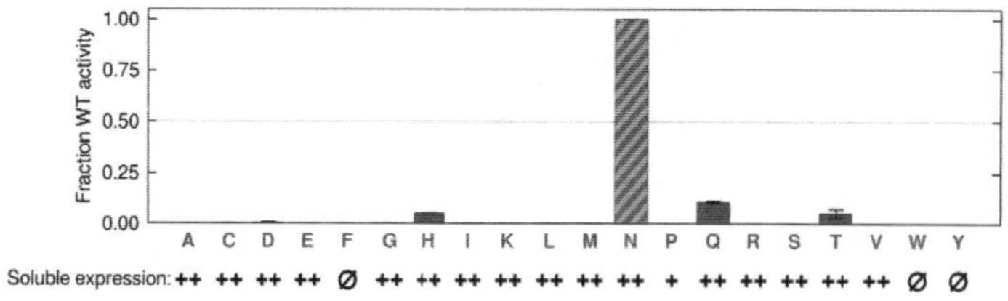

14

Global vs local views of sequence space

- Bgl3 is a member of glycoside hydrolase family 1 (GH1)
  - large and diverse enzyme family
  - members differ by hundreds of mutations
- Global view
  - GH1 sequence conservation score (1300 family members)

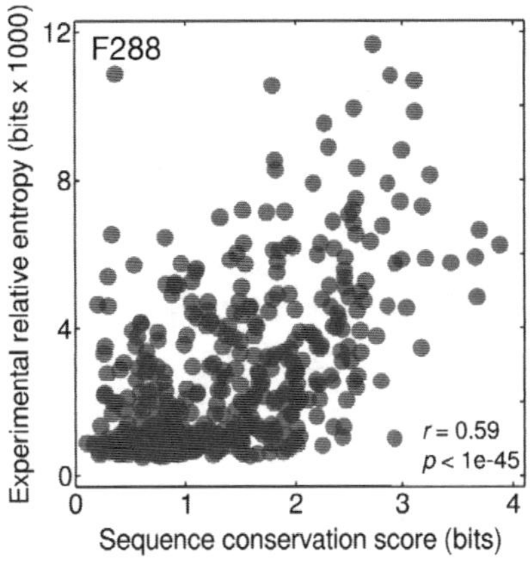

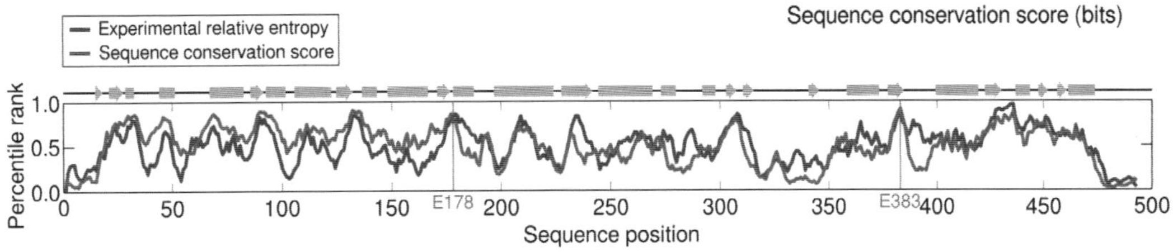

15

Structural diversity may explain outlying positions

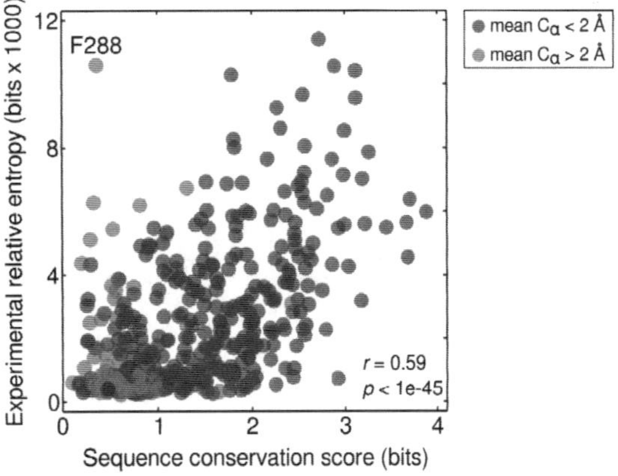

- Bgl3 may have evolved unique structural motifs that constrain it's mutational tolerance relative to other GH1 members
- Local view of sequence space highlights individuality of members
- Identifies interactions overlooked by large alignments of homologous sequences

16

High-temperature screening of beta-glucosidase library

- Droplet-based microfluidics allows precise control of screening conditions such as time, temperature, and concentration
- Screening under altered conditions enriches for variants with enhanced properties
- Modified microfluidic workflow to include a 10 minute 65 °C heat challenge
  - screened ten million enzyme variants
  - sorted two million active variants (20% functional)

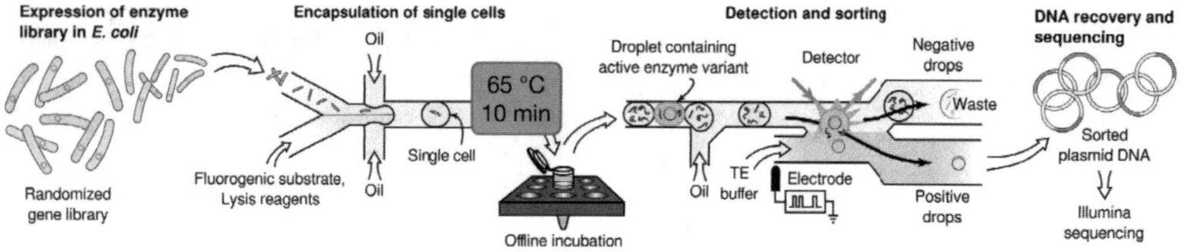

17

High-temperature mutational enrichment

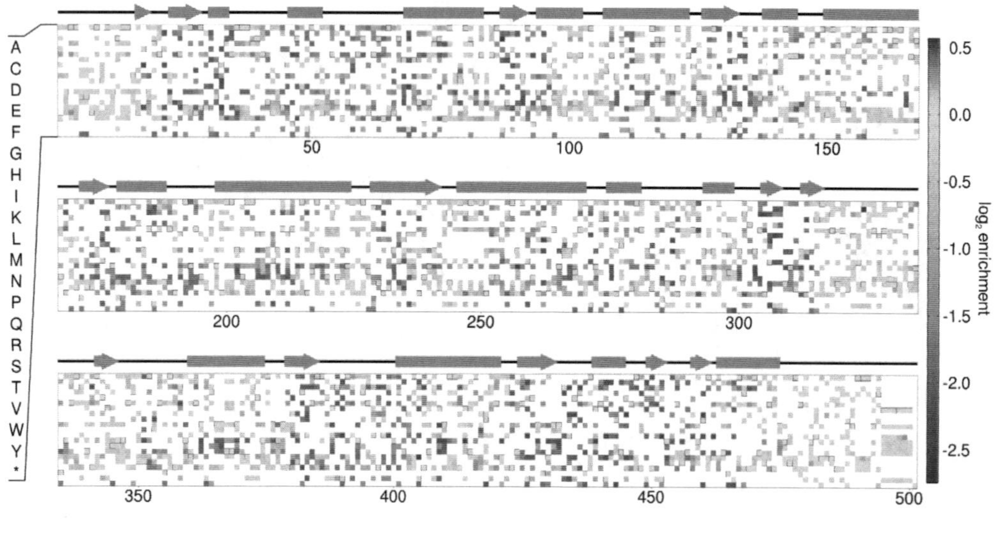

18

High-temperature screening enriches for stabilizing mutations

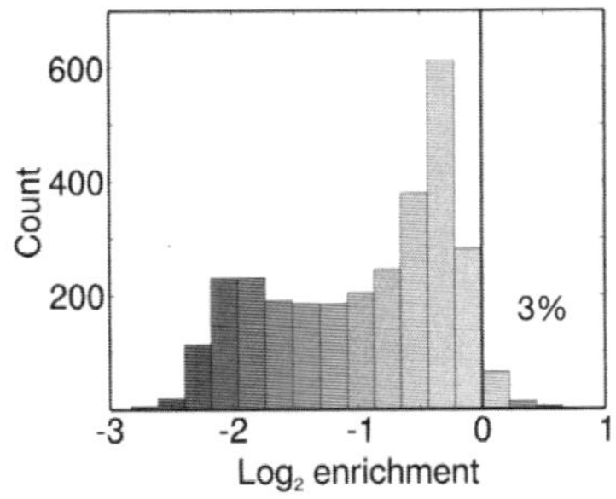

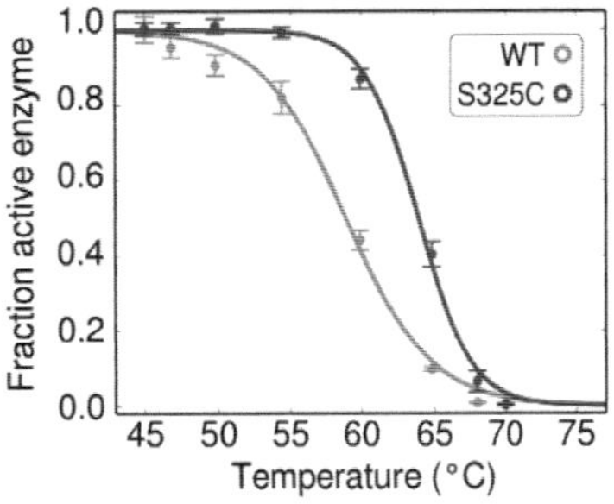

mutation	$\log_2$ enrichment	T_{50}	ΔT_{50}
WT	0.00	58.8	-
Y50F	0.12	60.4	1.6
A150D	0.16	59.3	0.5
H323D	0.25	59.9	1.1
S325C	0.57	64.1	5.3
P335A	0.32	59.1	0.3

19

High-throughput mapping of enzyme sequence-function relationships

- Droplet-based microfluidic screening enables high-throughput mapping of ***enzyme*** sequence-function relationships
  - readily adapted to enzymes with fluorescence-based activity assays
- Explore the molecular basis of enzyme function in an unbiased manner
  - discover new functional sites
  - local view of sequence space
- Flexible platform for assaying enzymes over a range of reaction conditions
  - high-temperature screening enriched for stabilizing mutations
  - other properties: k_{cat}, K_m, ionic liquid tolerance

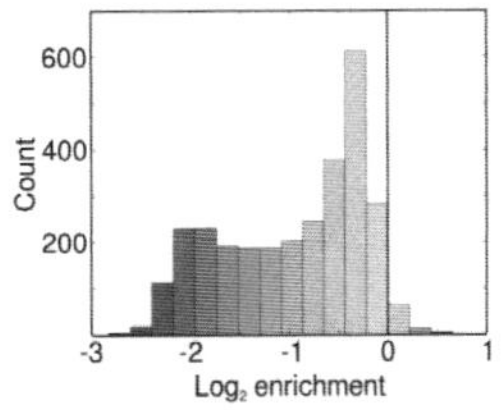

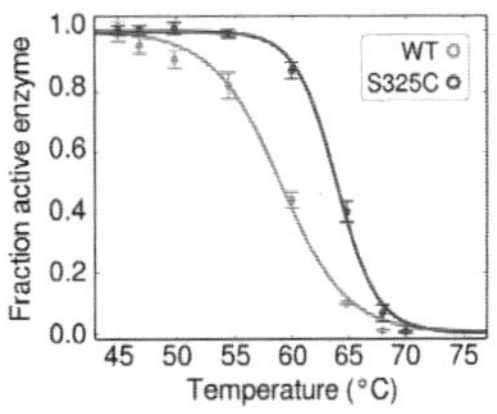

20

Acknowledgments

UCSF

Adam Abate
Tanja Kortemme
Tuan Tran
Abate laboratory
Kortemme laboratory

jbei
Joint BioEnergy Institute

Blake Simmons
Rich Heins

Funding: NIH, UCOP, BWF

Engineering Picomolar Affinity into a Rationally Identified 5 Kda Scaffold for Tumor Targeting

Kruziki, M. A., University of Wisconsin Madison
Zhou, H.,
Holec, P., University of Minnesota
Hackel, B. J., University of Minnesota
Originally presented on: 1/14/2015 0:00:00 - 0:25:00

Engineering Picomolar Affinity into a Rationally Identified 5 kDa Scaffold for Tumor Targeting

Max Kruziki, Hong Zhou, Patrick Holec, and Benjamin Hackel, University of Minnesota

Small protein scaffolds, relative to antibodies, can provide superior physiological distribution, stability, production, and ease of conjugation. Yet while many scaffolds have been tested to generate synthetic ligands, the causes of the varying degrees of success or failure have not been systematically explored. We merged computational and experimental tools to elucidate the biophysical factors that dictate evolutionary potential to empower rational identification or design of protein scaffolds. We have identified the 45-amino acid Gp2 domain as an effective protein scaffold to engineer stable, picomolar affinity ligands.

We developed and implemented a computational algorithm to efficiently evaluate naturally occurring protein domains on numerous metrics with potential impact on evolution of binding function. In one implementation, we identified all domains in the Protein Data Bank with two accessible loops (diversified paratope) and scored them on the size, shape, orientation, and target accessibility of their paratopes as well as protein size, stability upon mutation, and dependence on disulfide bonds and/or cofactors. Machine learning was used to preliminarily weigh each metric based on historical scaffold performance.

The top scaffold was the 45-residue T7 phage gene protein 2 with a single alpha helix opposite a beta sheet with two adjacent loops amenable to mutation. We diversified the twelve loop amino acids using complementarity bias and length diversity. The library of 10^8 mutants was displayed on the yeast surface and screened for binders to several protein targets. Novel ligand discovery and directed evolution yielded binders with affinities as strong as 200Â±100 pM and no observable nonspecific binding. Circular dichroism showed secondary structure comparable to the wild-type protein and T_m values from 70Â±4Â°C to 80Â±1Â°C for evolved ligands relative to 67Â±4Â° for wild-type. We evolved binders to epidermal growth factor receptor and MET that effectively label human tumor cells. Ligands are readily conjugated with diagnostic moieties. Updated progress on tumor targeting in murine models â?? via fluorescence tomography and positron emission tomography â?? will be presented as well as scaffold and evolutionary optimization.

More broadly, libraries from the top 30 scaffolds were created for comparative analysis of the ability to evolve specific, high affinity binding to numerous targets. We will discuss the relative efficacy of each scaffold as well as the scaffold design metrics that correlate most strongly with performance. Implications for future scaffold design, including synthetic protein topologies, will be addressed as well as other designs.

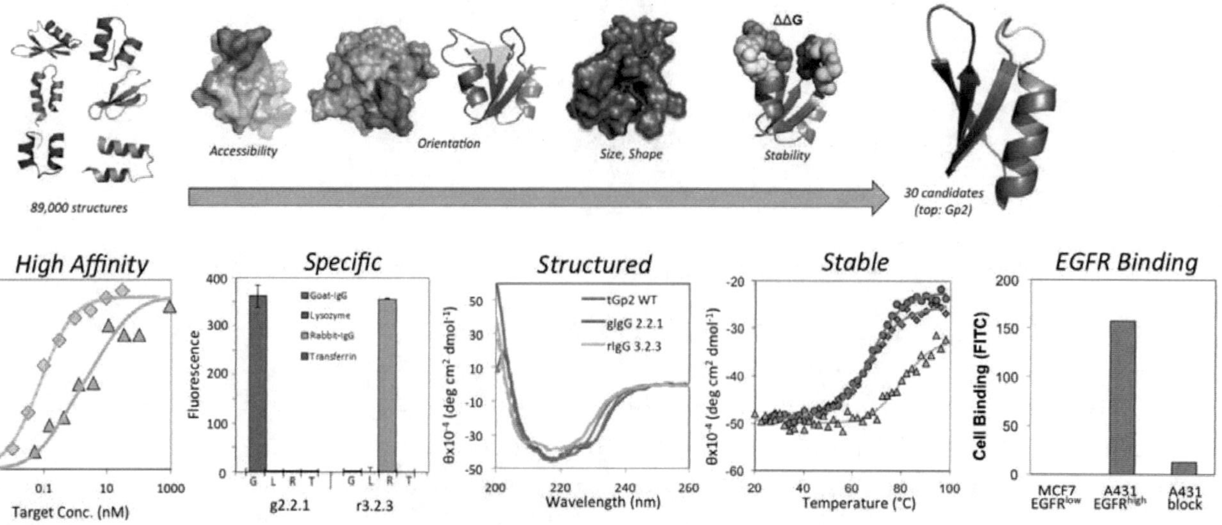

Scaffold identification. All structures within the Protein Data Bank were evaluated for the size (accessible surface area of the average mutated clone), shape (perimeter:area ratio and absence of intervening non-diversified framework), orientation (separation and simultaneous accessibility), and target accessibility of their paratopes as well as protein size, stability upon mutation, and dependence on disulfide bonds and/or cofactors. The top scoring scaffold from the preliminary analysis was T7 phage gene protein 2 (Gp2). Directed evolution of binding ligands from a diversified Gp2 library yielded clones with high affinity (≥200 pM), specific binding as evaluated on the yeast surface as well as with soluble ligands, maintained secondary structure, reversible thermal denaturation, and high stability as evaluated by circular dichroism, and target-specific binding in human cell culture.

Ultra-High Affinity Engineered Protein Therapeutics for Treating Metastatic Disease

Cochran, J. R., Stanford University

Originally presented on: 1/14/2015 0:00:00 - 0:25:00

We create engineered growth factor ligands and receptors that possess altered biochemical and biophysical properties with several goals in mind. For example, these engineered proteins are being used as tools for studying the relationship between properties such as binding kinetics and affinity and biological effects including cell signaling, proliferation, and migration. We also use growth factor ligands and receptors as starting points for the development of engineered protein therapeutics. Recent results on an engineered soluble receptor 'decoy', based on the Axl receptor tyrosine kinase extracellular domain, will be presented. Aberrant signaling through the Axl receptor tyrosine kinase has been associated with a myriad of human diseases, most notably metastatic cancer, identifying Axl and its ligand Gas6 as important therapeutic targets. Using rational and combinatorial approaches, we engineered an Axl 'decoy receptor' that binds Gas6 with high affinity and inhibits its function, offering an alternative approach from drug discovery efforts that directly target Axl. Four mutations within this high-affinity Axl variant caused structural alterations in side chains across the Gas6-Axl binding interface, stabilizing a conformational change on Gas6. When reformatted as an Fc fusion, the engineered decoy receptor bound Gas6 with femtomolar affinity, an 80-fold improvement compared to binding of the wild-type Axl receptor, allowing effective sequestration of Gas6 and specific abrogation of Axl signaling. Moreover, increased Gas6 binding affinity was critical and correlative with the ability of decoy receptors to potently inhibit metastasis and disease progression *in vivo*.

Modular Receptor Engineering for Programming Cell-Based Therapies to Interface with Host Physiology

Daringer, N., Northwestern University
Dudek, R. M., Northwestern University
Schwarz, K. A., Chemical and Biological Engineering, Chemistry of Life Processes Institute, R.H. Lurie Comprehensive Cancer Center, Northwestern University
Leonard, J. N., Northwestern University
Originally presented on: 1/14/2015 0:25:00 - 0:50:00

Engineered cell-based therapies represent an emerging frontier in medicine, and the promise of this approach has recently been demonstrated through curative eradication of B cell malignancy from a growing list of patients. However, extending this success to other types of cancer and to other applications in health and medicine will require new tools enabling bioengineers to custom program the manner in which engineered cells interface with host physiology. Until recently, we lacked the ability to construct customizable cell-based "devices" that detect and respond to exclusively extracellular cues, which include many species of biological relevance including cytokines, chemokines, cell-surface antigens, and pathogens. To meet this need, we developed a Modular Extracellular Sensing Architecture. This protein engineering platform comprises a self-contained receptor and signal transduction system, wherein ligand binding at the cell surface is transduced into a change in intracellular state (i.e., induction of gene expression) without requiring involvement of native intracellular signaling mediators, and therefore without being subject to native mechanisms regulating signal transduction. Here, we report our expansion upon this platform to enable the engineering of receptors that recognize essentially any external ligand and regulate essentially any native gene. We have developed general and rapidly employable engineering approaches for incorporating modular ligand binding domains (including both single-chain variable fragments, scFv, and single domain Nanobodies) to create new functional receptors. We have also engineered MESA receptors to generate "output" via Cas9-based gene regulation systems. Cas9-based transcriptional activators and repressors are easily targeted to regulate either native genes or engineered transgenes by co-expression with single-guide RNA (sgRNA) tailored to target a specific gene or set of genes. Finally, we have demonstrated that these approaches may be combined to functionally rewire mammalian cell input-output behavior. Ultimately, this capability provides a powerful tool for experimental systems biology, and in the context of medicine, enables the rapid evaluation of multiple potential strategies for functionally rewiring multicellular networks to achieve a therapeutic objective. We will present our application of this approach to modulating immune function, toward the goal of therapeutically modulating local immune states to overcome persistent barriers to immunotherapy of cancer.

Non-Native Small-Molecule Biosensors for Screening and Selection of High-Performance Cell Factories

Jensen, M. K., Technical University of Denmark
Skjødt, M.,
Zhang, J., Technical University of Denmark
Rajkumar, A. S.,
Keasling, J. D., Lawrence Berkeley National Laboratory
Originally presented on: 1/14/2015 0:00:00 - 0:25:00

Our research addresses the current needs for high-throughput screening of candidate chemical production from engineered microorganisms. Using non-native *in vivo* protein-based sensors we facilitate the screening and detection of microbial cell factories with improved production titers. The protein-based biosensors couple intra- or extracellular levels of candidate chemicals with growth, or by fluorescent reporter gene activity. Generally, we aim for simple, orthogonal and specific biosensors, applicable to microorganisms such as yeast and bacteria. For this session we will present the design optimization of both biosensor expression level, dynamic range and the design of synthetic promoters for orthogonal transcriptional regulation of growth and reporter gene outputs upon the intra- or extracellular accumulation of candidate chemicals.

Designing Microbes for Executive Function

Bentley, W. E., University of Maryland, College Park
Payne, G. F., University of Maryland, College Park
Terrell, J., University of Maryland
Wu, H. C., University of Maryland
Tsao, C. Y.,
Gordonov, T.,
Zargar, A.,
Quan, D.,
Originally presented on: 1/14/2015 0:25:00 - 0:50:00

Signal molecules play an important role in determining biological function. Signaling among bacteria is mediated by small molecules that are produced, secreted, accumulated, perceived and transduced in a variety of manners that span biophysical and genetic regulatory domains. Often the signal molecule's structure is determinant in defining function. Self-reporting number density is referred to as quorum sensing is such an example; its use enables phenotypic noise reduction for specific cell subpopulations in natural environments. Sometimes, the signal transduction process confers bacteria with the ability to establish their own virulence. Our studies have contributed to the discovery of QS, its components, and its utility as a source for genetic "cassettes" that can be imparted to cells to reprogram their function. Quorum sensing is also an outstanding "testbed" for connecting biological signaling phenomena with microfabricated devices. The communication systems are divergent: biology "communicates" with small molecules and ions, whereas devices communicate with electrons. Understanding the connectivity and translating the systems for two way communication will open many new avenues for understanding biological systems and manipulating their function. That is, a second focus of our work is on the "biofabrication" of devices that enable recapitulation and recording of biological signaling events by transducing them to electrical signals read via microfabricated chips. We are developing an entirely new toolbox for device fabrication that builds on simple, low-cost, and reagentless assembly techniques.

Our presentation will discuss new concepts that build on synthetic biology and biofabrication as a means to interrogate information embedded in collections of small molecules. Cell constructs serve as filters, interpreters, and information amplifiers for transmission of information to abiotic signal amplifiers and finally to end users. The synergy between synthetic biology and nanomaterials will be discussed as a means for transmitting and receiving information via molecular networks.

Engineering Orthogonal Translation Systems

Jewett, M. C., Northwestern University

Originally presented on: 1/14/2015 0:50:00 - 1:15:00

The translation apparatus is the cell's factory for protein synthesis, stitching amino acid substrates into sequence-defined polymers (proteins) from a defined template. With protein elongation rates of up to 20 amino acids per second and an accuracy of > 99.9%, prokaryotic translation is faster and more accurate than many enzymatic conversions that require only a single biocatalyst and a single reactant. This extraordinary synthetic capability, combined with the ability to produce long polymers sustained by adapter molecules for genetic encodability, has driven extensive efforts to harness the translation apparatus for novel functions. Pioneering efforts have demonstrated that it is possible to genetically encode more than the 20 natural amino acids and that this expansion can be a powerful tool. For example, the site-specific introduction of non-standard amino acids (nsAAs) into proteins has generated novel biological insights and applications that would be difficult, if not impossible, to achieve by other means. Despite establishment of ~100 orthogonal translation systems (OTSs) for nsAA incorporation and a rapidly growing number of cases showing their utility, building designer OTSs with high activity and specificity remains a significant challenge. Common challenges include poor efficiency of engineered OTS components, compatible ribosomes and elongation factors, and constraints arising from the fact that microbial growth and adaptation objectives are often opposed to the engineer's process objectives. In this presentation, I will discuss our efforts to address these challenges by engineering synthetic ribosomes and using cell-free systems for the synthesis of proteins containing multiple nsAAs. Our work is enabling a deeper understanding of why nature's designs work the way they do and opening new frontiers for harnessing a dramatically expanded genetic code for manufacturing novel therapeutics, synthesizing genetically-encoded materials, and elucidating fundamental biological insights.

Microbial Production of Curcuminoids

Funa, N., University of Shizuoka
Originally presented on: 1/14/2015 1:15:00 - 1:40:00

Curcuminoids, which are produced specifically by Zingiberales, have been used as food additives because of their aromatic, stimulant, and coloring properties and as traditional Asian medicines because of their anti-tumor, antioxidant, and hepatoprotective activities. We have discovered a novel type III polyketide synthase catalyzing the synthesis of curcuminoids from *Oryza sativa*. Curcuminoid synthase synthesizes bisdemethoxycurcumin via a unique mechanism from two 4-coumaroyl-CoAs and one malonyl-CoA. The reaction begins with the thioesterification of the thiol moiety of Cys174 by a starter molecule, 4-coumaroyl-CoA. Decarboxylative condensation of the first extender substrate, malonyl-CoA, onto the thioester of 4-coumarate results in the formation of a diketide-CoA intermediate. Subsequent hydrolysis of the intermediate yields a beta keto acid, which in turn acts as the second extender substrate. The beta keto acid is then joined to the Cys174-bound 4-coumarate by decarboxylative condensation to form bisdemethoxycurcumin. This reaction violates the traditional head-to-tail model of polyketide assembly; the growing diketide intermediate is hydrolyzed to a beta keto acid, which subsequently serves as the second extender to form curcuminoids.

An artificial curcuminoid biosynthetic pathway, including reactions of 4-coumarate:CoA ligase and curcuminoid synthase was constructed in *Escherichia coli* for the production of curcuminoids. Cultivation of the recombinant *E. coli* cells led to production of bisdemethoxycurcumin, dicinnamoylmethane and cinnamoyl-*p*-coumaroylmethane from exogenously supplemented phenylpropanoid acids, such as *p*-coumaric acid, cinnamic acid and ferulic acid.

Malonyl-CoA is an important intermediate in the pathway of the *de novo* fatty acid biosynthesis and fatty acid elongation. The rate of fatty acid synthesis is strictly controlled by the regulation of acetyl-CoA carboxylase, which catalyzes carboxylation of acetyl-CoA to yield malonyl-CoA. In E. coli, it has been shown that the cellular availability of malonyl-CoA is limited due to its direct association with cell growth and the biosynthesis of fatty acids. Besides being a rate-limiting compound of the central metabolic pathway, malonyl-CoA is a major building block of secondary metabolites, such as polyketides, in plants and microbes. For example, naringenin chalcone, a key intermediate of the biosynthetic pathway of flavonoids, is synthesized from p-coumaroyl-CoA and three molecule of malonyl-CoA by chalcone synthase, which is a plant specific type III polyketide synthase. In addition, RppA, a type III polyketide synthase from Streptomyces griseus, catalyzes the condensation of 5 malonyl-CoAs to synthesize 1,3,6,8-tetrahydroxynapthalene, which is an intermediate of DHN melanin. *E. coli* is a genetically tractable microorganism that can express multiple genes simultaneous by using expression vectors with different replication origins and selectable markers. Moreover, the fermentation process of *E. coli* is manageable and reproducible. Although, *E.coli* has been used for a host for the large-scale production of natural products, the reconstitution of the plant biosynthetic pathways with type III polyketide synthases in *E. coli* often resulted in a low production rate. To overcome the obstacle impeding the use of type III polyketide synthase for the large-scale production of polyketides, several approaches have been shown to improve the cellular concentration of malonyl-CoA. The activation of the pathways leading to the synthesis of malonyl-CoA and the elimination of the competing pathways reducing the precursors of malonyl-CoA are the major targets of the metabolic engineering. Inactivation of fatty acid biosynthesis will be presented.

Bias and Error Correction in Antibody Repertoire Sequencing for More Accurate Vaccine Profiling

Khan, T., ETH Zürich
Reddy, S. T., University of Texas Austin

Recent advances in Next-generation DNA sequencing (NGS) technology have enabled high-throughput systems analysis of immunological responses. NGS of antibody repertoires has emerged as a method to support or augment existing immunological technologies such as screening (e.g. hybridoma, recombinant surface display, etc), antibody measurements (e.g. titers), and cellular characterization (e.g. ELISPOT, cell culture supernatant cytokine quantification, and flow cytometry). One potential pitfall of antibody NGS is the lack of uncertainty when quantifying the relative abundance between antibody clones, which compromises the quantitative accuracy of repertoire data. This uncertainty stems from systematic and stochastic biases introduced during sample preparation (i.e. PCR). This is especially problematic for samples derived from sources that currently rely on primer sets, where slight differences in individual primer annealing temperatures manifest into large systematic biases in clonal relative abundance. These primer sets used for NGS of antibody variable regions from mice and humans have relied on multiplex primers composing up to 150 different individual primers to compensate for the high diversity of V genes. Therefore, we have developed an experimental-bioinformatic approach to ensure accurate quantitative assessment of clonal frequencies based on molecular transcript barcoding (MTB). MTB relies on generating cDNA with unique identifiers, incorporated by gene specific primers in the form of oligonucleotides with specified degenerate regions, to enable bias calibration data. During method development, we employed qPCR assessment to minimize bias effects due to over amplification. After PCR amplification, we then collect MTB NGS sequencing data on the Illumina MiSeq platform enabling read lengths of 2x300. Identified biases are then able to be corrected using bioinformatics approaches to produce more accurate information about clonal relative abundance. In addition to correction of clonal frequencies, MTB enables removal of sequencing error and PCR error, thus ensuring high quality data and distinction of synthetic mutations from true biological mutations (somatic hypermutations). Furthermore, by introducing unique identifiers for each transcript, we are also developing a process to recover bioinformatically relevant genes using primers targeting those unique identifiers (dial-out PCR). By providing more accurate antibody amplicon NGS information, we are using various statistical approaches to characterize the degree of skewed clonal distributions post-immunization in mice (e.g. Shannon entropy). These types of analyses will lead to improved strategies for creating better quantitative vaccine response characterization methods. We believe these strategies will be able to support quantitative serum based epitope mapping characterization, and as a single method could be used to evaluate responses to all antigens and pathogens. This work will lead to increased speed and efficiency for profiling of vaccine-induced immune responses.

Studies on the Influence of Different Metabolic Uncouplers on the Biodegradation of Toluene in a Differential Biofilter Reactor

Detchanamurthy, S., Sri Venkateswara College of Engineering
Gostomski, P., University of Canterbury

One of the biggest challenges in a traditional biofilter is to overcome the low volumetric degradation rate, which often makes the footprint excessive. The volumetric degradation rate or elimination capacity (EC) is directly influenced by the specific biodegradation rate of the microorganism involved in degrading the pollutants. Application of metabolic uncouplers in biofiltration could improve the biodegradation rate of microorganisms, which in turn could increase the EC. The addition of metabolic uncouplers to the growth system decreases the biomass growth whereas in energy-excess, non-growth systems like biofiltration, it is expected to increase the specific substrate uptake rate since the substrate requirement for maintenance energy should increase, which in turn should increase the EC. Seven potential metabolic uncouplers were screened in batch serum bottles and subsequently tested in a continuous biofilter reactor with soil as the biofilter medium. The metabolic uncouplers tested were benzoic acid, carbonylcyanide p-trifluoromethoxy phenylhydrazone (FCCP), carbonylcyanide m-chloromethoxy phenylhydrazone (CCCP), pentachlorophenol (PCP), malonic acid, 2, 4, 6-trichlorophenol (TCP) and m-chlorophenol (mCP). Only PCP and 2, 4, 6-TCP increased the toluene degradation rate significantly. PCP increased the toluene degradation rate by 35% at 140 μM, whereas 4051 μM TCP increased the rate by 18%. FCCP did not significantly affect the degradation rate and the other metabolic uncouplers decreased the degradation rate. Five toluene degraders were isolated from soil subjected to toluene and were identified using 16s rDNA/18s rDNA analysis. Out of five, two potential toluene degraders, *Stenotrophomonas maltophilia* and *Pseudomonas putida* were used to develop a biofilm reactor. PCP, TCP and CCCP were tested in the biofilm reactors and found that PCP increased the surface elimination capacity (SEC) by 85% at 140 μM in *S. maltophilia* biofilm reactor and CCCP increased the SEC by 27% at 1 μM in *P. putida* biofilm reactor.

A Biobased Approach to Mechanically Tunable Polyester

Zhang, K., University of Minnesota

While holding great promise as sustainable and biodegradable, current biobased polyesters encounter significant difficulties in commercialization because of their brittleness and rigidity. This characteristic precludes their use in products required to be soft, elastic, or flexible. Recently we have developed a biobased approach to manufacturing a branched lactone, β-methyl-δ-valerolactone (MVL), and demonstrated its potential as the first scalable rubbery polyester. We designed an artificial metabolic pathway to MVL by extending the mevalonate pathway: introduction of fungal siderophore proteins to convert mevalonate into anhydromevalonolactone; and reduction of anhydromevalonolactone to MVL by enoate reductases. The total biosynthetic pathway produced 300 mg/L of MVL directly from glucose in shake flask experiments. While directed evolution is ongoing to optimize the MVL pathway, as an alternative solution we developed a semisynthetic approach for the immediate commercial-scale production of MVL. In this approach we introduced the archaeal mevalonate pathway into production host *E. coli* and engineered the pathway to produce ~90 g/L mevalonate in a bioreactor. We then performed two simple chemical reactions, dehydration & hydrogenation, to produce MVL from mevalonate. The overall purification and reaction efficiency reached 90% . With the biobased MVL, we generated an amorphous rubbery polymer with a T_g at -51°C, which can be used to prepare mechanically tunable polyesters for packaging, thermoplastic elastomers, agricultural mulch films, chewing gums and pressure sensitive adhesives.

Tailor Designed Peptides Bio-Inspired from Transmembrane Proteins for the Cosmetic Industry

Aguilera Segura, S. M., Grupo de Diseño de Productos y Procesos (GDPP) Universidad de los Andes
Nuñez Velez, V. L.,
Alvarez Solano, O. A., Universidad de los Andes
Achenie, L., Virginia Tech
Torres Saez, R.,
González Barrios, A. F., Universidad de los Andes

Surfactants are amphiphilic compounds with the ability to radically alter surface and interfacial properties. These properties are highly desired in food, health, personal care, and oil industries in which there are several processes related with stabilization of emulsions. Several efforts have been invested on the search of environmentally benign products, and there is a need for a strategy to identify biosurfactants that can replace conventional ones. The aim of this work is to evaluate transmembrane protein-engineered peptides via Free Energy (FE) Molecular Dynamics (MD) and to compare them against the already experimentally tested $OmpA_{325}$. Sixteen protein-engineered peptides were selected from the sequence of $OmpA_{325}$ as potential biosurfactants based on its hydrophobic profile. The FE-methodology comprised a series of MD simulations of the path between two states of interest, discretized by several intermediate ones, and coupling van der Waals and Coulomb interactions separately. After FE-MD, an analysis from the simulations to obtain a statistical estimate for the free energy with the Bennett's acceptance ratio (BAR) method was performed. Analysis of the 5-ns MD trajectories showed the potential ability of all peptides to increase the system's stability. Negative ΔG_{solv} was achieved in all cases, with significant phase-space overlap and nearly-zero standard deviations. Then they were scored and ranked by taking into account the ΔG_{solv}, solvent accessible surface area (SASA), molecular weight (MW), and $\Delta G_{Coul}/\Delta G_{vdW}$ ratio for final experimental validation utilizing water-dodecane emulsions with each synthetized peptided where stability and drop size distribution were evaluated and compared with the list provided by MD simulations.

Energizing Yeast Cell-Free Protein Synthesis with Glucose Metabolism

Stark, J. C., Northwestern University
Anderson, M., Northwestern University
Hodgman, C. E., Northwestern University
Jewett, M. C., Northwestern University

Energizing yeast cell free protein synthesis with glucose metabolism

Jessica C. Stark, Mark Anderson, C. Eric Hodgman, Michael C. Jewett Abstract
Cell-free protein synthesis (CFPS) is a powerful technology with a growing number of applications, including high-throughput synthesis of protein libraries and characterization of genetic components. There currently exist high-yielding *Escherichia coli* CFPS platforms that avoid the use of expensive high-energy phosphate compounds. However, to date, eukaryotic CFPS systems generally utilize such compounds to regenerate the adenosine triphosphate (ATP) necessary to drive protein synthesis. Expensive reagent costs have prevented the widespread use and practical implementation of eukaryotic CFPS technology. In this study, we report the development of the first ever, to our knowledge, eukaryotic CFPS system powered by natural energy metabolism of non-phosphorylated energy substrates. To achieve this, we screened six different glycolytic intermediates for their ability to regenerate ATP and fuel protein synthesis in a *Saccharomyces cerevisiae* crude extract CFPS platform. We observed the synthesis of 1.05 Â± 0.12 Âµg mL^{-1} luciferase when using 16mM glucose as a secondary energy substrate and demonstrate that glycolysis is active by quantifying the production of ethanol during the reaction. With the addition of potassium phosphate, our yields using glucose increased approximately 3.5-fold to 3.64 Â± 0.35 Âµg mL^{-1}. Although synthesis yields on a gram per liter basis remain lower than the CrP/CrK system previously developed, the cost-to-yield ratio is comparable. These promising developments provide the foundation for further development of a cost-effective eukaryotic CFPS system. Further, this work provides the first evidence that glycolytic metabolism is active in eukaryotic crude extract CFPS platforms. This demonstration may lead the way to development of cost-effective eukaryotic CFPS platforms from multiple host organisms for high throughput protein expression, synthetic biology, proteomics, and structural genomics applications.

Investigation of Metabolic Capabilities of Recombinant Lactococcus Lactis for Production of Hyaluronic Acid Using Constraint Based Genome Scale Models

Badri, A., Indian Institute of Technology, Madras
Raman, K., Indian Institute of Technology, Madras
Jayaraman, G., Indian Institute of Technology,Madras

Background:

Hyaluronic acid (HA), a naturally occurring polysaccharide, is one of the chief components of the extracellular matrix in eukaryotes and the extracellular capsule in *Streptococci*. It is a linear non-sulphated glucosaminoglycan that contains alternating units of Î^{2}-1,3-N-acetyl glucosamine and Î^{2}-1,4-glucuronic acid. Its excellent water binding capacity paves way for its wide application in various fields including ophthalmic surgery, viscoelastic supplementation in arthritis and drug delivery scaffolds.

At first, HA was primarily extracted from rooster comb. However, the convoluted protocol of extraction and the chances of cross-species infection from avian viruses discouraged its application in biomedical areas. On the other hand, microbial sources like *Streptococci,* which are natural producers of HA, were more attractive as far as industrial production was concerned. However, the pathogenicity of *Streptococci* limited their use in biomedical applications. Hence, several alternatives are being developed by incorporation of HA synthesizing ability in â??saferâ?□ organisms such as *Bacillus subtilis*, *Lactococcus lactis*, *Agrobacterium sp.* and *Escherichia coli.*

Our research group focuses on *Lactococcus lactis*, a lactic acid bacterium, for production of HA. Currently, *Lactococci* are seen as ideal avenues for metabolic engineering due to the simplicity of their metabolism. They are widely used in many food fermentation industries and are classified as GRAS (Generally Regarded As Safe), making them more favourable for industrial production and medical applications.

A major issue concerning the recombinant microorganisms developed for HA production is the low HA yield and molecular weight when compared to *Streptococci.* Several research groups worldwide are studying the effect of process parameters and media constituents on HA yield and molecular weight to optimise the production strategy. However, to successfully engineer the cell to harness its maximum capabilities, it is also important to understand its metabolism in detail.

Redirection of fluxes towards HA pathway by modifying the metabolic network of the organism is another approach to contrive the cell for improved HA production. The general view is that flux towards HA biosynthesis pathway competes with glucose metabolism fluxes (glycolysis, PPP etc.) for glucose. Under normal culture conditions, it is known that the

glycolysis pathway is the major glucose consuming pathway. Hence glycolytic fluxes are central in the understanding of HA biosynthesis. The precursors of HA are also formed from glycolytic intermediates (glucose-6-phosphate and fructose-6-phosphate). On this note, there are successful studies of enhancement of HA production upon partial inhibition of glycolysis.

We here use a genome-scale metabolic model to analyse fluxes of *Lactococcus lactis* for producing HA *in silico*. Constraint-based modelling techniques such as Flux Balance Analysis (FBA) have been proven to be an effective tool to interrogate metabolic networks and understand the metabolic capabilities of organisms. Apart from predicting a theoretical maximum HA yield, the model also enables the analysis of metabolic fluxes under different conditions. This would lead towards a better understanding of the obscure factors playing a role in production of HA. Further, we will also identify strategies for overproducing HA, through gene deletions as well as over-expression of critical genes.

Methods:

The most recent version of the genome-scale metabolic model of *Lactococcus lactis* MG1363 was published in 2013 (Flahaut *et al.*, Appl Microbiol Biotechnol. 97(19):8729). The metabolic model involves a complete catalogue of all (known) reactions taking place in the organism, complete with stoichiometry and enzyme information. Based on constraints on the flux values of each reaction and an overall cellular objective, FBA involves the formulation of a linear programming problem, which is solved to obtain a set of permissible fluxes that will achieve a given objective (e.g. maximum cellular growth). Since the original model did not specify precise upper and lower bounds for the reaction fluxes, we discarded the flux bounds in the original model and replaced them with some extracellular bounds determined experimentally for the current system of recombinant *Lactococcus lactis* capable of producing HA. Following this, we evaluated the ability of the model in reproducing trends observed in laboratory experiments. We also systematically analysed the model for possible bottlenecks and over-expression targets that increase flux towards HA synthesis. The aim was to analyse if there are any important reactions that affect flux towards HA production in the entire reactome of the cell, apart from the already known pathway reactions. Using a technique called FSEOF (Flux

Scanning Based on Enforced Objective Flux), the set of reactions whose flux increased upon forced increased of HA flux were collected. FSEOF enforces objective flux to various fractions of maximum HA flux (say 0.2, 0.3, â?¦, 0.8) and performs an FBA to maximise biomass, for each case. Across these simulations, FSEOF identifies reactions that carry an increased flux, as HA production increases. The premise is that these reactions, which uniformly show an increase in flux, with increase in HA production, may be potential over-expression targets. However, there are chances that some of these increased co-incidentally along with increase in HA flux and these need to be weeded out to get the actual over-expression target list.

Results:

The model was able to reproduce the experimentally observed trends of the effect of certain factors on HA. For example, the model predicted that an increase in glutamine uptake flux, a decrease in lactate production flux, an increase in acetate production flux and a decrease in flux towards fructose-6-phosphate individually increased HA formation flux. The stoichiometric effect of currency metabolites on HA was also checked. Increase in ATP formation fluxes in the system showed increase in HA production flux and vice versa.

FSEOF analysis identified several amino acid biosynthesis reactions and nucleotide metabolism reactions apart from HA pathway reactions. The involvement of glutamine in HA pathway could be a possible reason why amino acid biosynthesis reactions were among the identified reactions. When we sought to identify why nucleotide metabolism reactions were in the list, we found that the model used the ribose sugar backbone from nucleotides as the carbon source to make the HA monomers under the conditions where the cell had to maximise HA production. Similarly, the actual effect of the amino acid biosynthesis reactions on HA production are to be evaluated too, to know if apart from increasing glutamine flux, any other amino acid is involved in an indirect way in the HA pathway. A few inorganic reactions such as phosphate uptake were also among the identified reactions. They need to be screened to check if they are co-incidentally increasing.

Our model, like most other genome-scale models, has certain limitations. This model is restricted to analysis of the stoichiometry and metabolic network of the cell alone. In reality, there are other important networks like regulatory networks and signalling networks that play along with the metabolic network in bringing about changes. Another limitation is that we can analyse

fluxes towards HA monomer production alone. The molecular weight cannot be determined from this model.

Overall, our simulations suggest alternate ways to improve the production of HA in *L. lactis*, through a combination of over-expression of genes in critical pathways such as nucleotide and amino acid metabolism apart from the usual targets in the HA pathway. Our model also provides several hypotheses for experimental verification, which we intend to carry out in the near future.

Development and in Vivo Testing of Antibodies Targeting Cancer-Associated Fibroblasts

Van Deventer, J. A., Massachusetts Institute of Technology
Rajan, S., University of Toronto
Sidhu, S., University of Toronto
Wittrup, K. D., Massachusetts Institute of Technology

An expanding body of work suggests that the tumor stroma plays an integral role in the initiation and development of epithelial cancers. Within the stroma, cancer-associated fibroblasts (CAFs) are involved in several functions that form key aspects of tumor progression, including immune suppression, metabolism, and invasion and metastasis. Despite the demonstrated importance of CAFs within the tumor microenvironment, targeting CAFs in a therapeutic context remains a challenge. The work presented here details the development of a suite of antibodies targeting multiple epitopes of fibroblast activation protein-alpha (FAP), a membrane-bound protease selectively overexpressed on CAFs in numerous epithelial cancers. Using yeast surface display with a synthetic antibody library, more than 50 unique antibodies recognizing FAP were identified, with close to 30 of these proteins recognizing both the human and murine forms of the antigen. In vitro characterizations using yeast-produced scFv-Fc forms of the binding candidates enabled the efficient identification of clones recognizing distinct epitopes of FAP. After affinity maturation, these candidates were converted to full-length IgGs and found to retain their favorable binding properties. These antibodies exhibited good biophysical characteristics and reasonable pharmacokinetic properties in mice, making them suitable for therapeutic studies. A combination immunotherapy involving one of the anti-FAP antibodies has resulted in potent control of tumor growth in a syngeneic mouse model of colon carcinoma. These results suggest that targeting FAP with an immunotherapeutic approach may serve as a general strategy for the treatment of solid tumors.

The Bio Phantom Tissue Model

Calderon, J.,
Calderon, C., University of Puerto Rico

Phantom tissues occupies a important role in phototherapeutic research. Current work have concentrated to deliver stable non-biodegradable speciments providing stability and repeatable optical properties. This research proposed a formulation woth the objective to offer vascularized charateristics. By introducing biologigal absorption agents and stable metalal based scattering agent over a silicon based , we demosntrated a model offering dynamic optical funcitons mimiking an epidermial organ. The material and procedures included samples synthesized according to ASTM 68 Standard , the based material is Dimethy-siloxane formulation (LT 334 from)Silicones- Inc) pigment produced from the liquir of a hexane extraction of the seed Bixa Orellana. Oprtical properties of sample irradiated with a 645 nm laser were produce using a Dynamic Spatial Freequency controller developed in our lab. A 23 joules/sq. cm irradiation showed modulated reduced scattering properties ranging from 3 mm to 6 mm. The bio phantom tissue demostrate a adequate model with resemblance to color changes due to light exposure.

In silico Prediction of Gene Deletion Targets in Escherichia coli for Enhanced Succinic Acid Production Using a Model-Guided Approach Under the Optflux Software Platform

Mienda, B. S.,
Shamsir, M. S.,

In silico prediction of gene deletion targets in Escherichia coli for enhanced succinic acid production using a model-guided approach under the OptFlux software platform

Bashir Sajo Mienda and Mohd Shahir Shamsir

Bioinformatics Research Group (BIRG), Biosciences & Health Sciences Department, Faculty of Biosciences and Medical Engineering, Universiti Teknologi Malaysia, Skudai 81310 Johor Bahru. Malaysia

Abstracts

The advent and proliferation of Escherichia coli genome scale metabolic models coupled with breakthroughs in computational tools, is revolutionizing the field of metabolic engineering and synthetic biology. E. coli has been metabolically engineered using a number of established experimental trial and error strategies, to enhance the production of succinic acid on glucose and glycerol substrates under anaerobic conditions. However, investigation of in silico prediction of gene knockout targets for increased succinic acid production, using a model-guided approach under the OptFlux software platform, is yet to be elucidated. Here, we show using the most recent metabolic reconstruction of E. coli iJ01366 model that in silico deletion of pyruvate formate lyase A (pflA/b0902) and lactate dehydrogenase A (ldhA/b1380) using the OptFlux software platform has been predicted to enhance succinic acid production using glucose and glycerol as solitary carbon sources. The mutant models constructed in this study exhibited higher succinate production, of 115% and 121% respectively, when compared to their parent models on glucose and glycerol. Accordingly, growth rates of the mutant models were maintained at 60% on glucose and 67% on glycerol compared to their respective wild-type models. We hypothesized that a significant succinate flux could be achieved by knocking out pflA/b0902 and ldhA/b1380 in E. coli using glycerol and glucose as substrates, and that the choice of substrate is critical to enhancing succinic acid production. The results further demonstrates that the OptFlux software using minimization of metabolic adjustment (MOMA) as the simulation algorithm holds great promise as a reference computational tool for in silico metabolic engineering applications, that can guide future experimental studies, which is not only limited to succinic acid production in E. coli, but also for other important chemical compounds produced by microbial cell factories.

Key words: In silico prediction, gene deletion, Escherichia coli, Model-guided approach, OptFlux succinic acid.

Bypassing Local Maxima in Protein Directed Evolution through Negative Selection

Steinberg, B., Johns Hopkins University
Ostermeier, M., Johns Hopkins University

Directed evolution strategies often use increasing stringency of positive selection to engineer improved proteins. However, proteins frequently exhibit a high degree of sign epistasis, implying a rugged fitness landscape containing many local maxima. Positive selection is constrained by this topology since the number of accessible evolutionary pathways that improve fitness is limited. Present approaches for bypassing local maxima, such as generation of library diversity through neutral drift or genetic recombination, rely largely on chance or require a connected pathway through equally fit sequences. Instead, we suggest that traversing a fitness valley using negative selection may efficaciously provide pathways to new maxima with higher fitness. As a model system, we compared four directed evolution strategies for increasing the antibiotic resistance conferred by *β-lactamase:* positive selection, neutral selection, negative selection, and oscillating selection. We used a synthetic gene circuit that kills cells possessing *β-lactamase* activity above or below a desired threshold to perform the neutral and negative selections. All four strategies ended with positive selection steps to climb to fitness peaks. Interestingly, the strategy that included negative selection yielded more highly active variants of *β-lactamase* than the other three selection strategies. We performed deep sequencing on the intermediate stages of the negative selection libraries to describe and reconstruct evolutionary pathways leading to these improved mutants. The ability of negative selection to provide access to novel fitness peaks has important implications for applied directed evolution as well as the natural evolutionary mechanisms, particularly of antibiotic resistance.

Controlling Cells through RNA Folding

Lucks, J. B., Cornell University

Cells have an amazing ability to process information, make decisions, and change their state in response to changing environments. This ability is encoded within the cellular DNA genome, which is converted into RNA and protein molecules through the basic processes of gene expression. Among the many functions these RNAs and proteins perform is regulating their own expression. In fact RNAs are now known to regulate almost all aspects of gene expression, and play central roles in controlling some of life's most basic processes.

As with many biomolecules, RNA function is intimately related to its structure. Being a single-stranded polymer of nucleotides, RNA can fold back on itself to form structures that enable certain RNAs to block or allow gene expression processes. Central questions in biology and bioengineering are then: How do RNAs fold inside cells?, and How can we engineer these folds to control gene expression? In this talk I will present our recent work including two new accomplishments that address both of these questions.

I will start by describing SHAPE-Seq, our technology that couples chemical probing with next-generation sequencing to characterize RNA structures on an 'omics' scale. I will also describe our recent innovations in using SHAPE-Seq to simultaneously characterize the structures and functions of RNA regulatory molecules as they exist *inside* cells. Using examples from two RNA translational repressors used in synthetic biology, I will show how this *in-cell* SHAPE-Seq technique is producing quantitative measures of cellular RNA structure at single-nucleotide resolution that we can now directly link to RNA regulatory function.

I will then present our work on engineering RNA molecules that control the basic process of RNA transcription, and on how our knowledge of the RNA structure/function relationship has allowed us to engineer an RNA regulatory mechanism not found in nature. This new mechanism, called Small Transcription Activating RNAs, or STARs, allows new types of RNA-based genetic circuits and logics to be built with intriguing properties. I will conclude with thoughts about how our work is leading us to ask deep scientific questions about the fundamental processes of RNA folding and function, and the role of RNAs in nature.

A Human Therapeutic Enzyme Specifically Sabotages Tumor Metabolism By an Engineered Cystine/Cysteine Degrading Activity

Cramer, S., The University of Texas at Austin
Saha, A.,
Tadi, S., The University of Texas at Austin
Tiziani, S., The University of Texas at Austin
Digiovanni, J.,
Stone, E. M., University of Texas at Austin
Georgiou, G., The University of Texas at Austin

Numerous studies have revealed that in contrast to normal tissue, high mortality cancers such as prostate carcinomas (PCa), glioblastomas, and small cell lung carcinomas are critically dependent on the uptake of the non-essential amino acid L-cysteine (L-Cys) and its oxidized form L-cystine (CSSC) for proliferation and survival. Cancer specific metabolic defects such as L-Cys dependence are unique therapeutic opportunities that can be exploited by introducing into circulation pharmacologically optimized enzymes that degrade the amino acid upon which the tumor critically relies. We hypothesized that enzyme-mediated systemic depletion of the serum L-Cys/CSSC pool would constitute a powerful and completely novel cancer therapeutic approach as a cancer monotherapy and as a mechanistically synergistic agent with existing chemotherapeutic modalities. Indeed in an initial protein engineering campaign exploring this hypothesis we *engineered and pharmacologically optimized a prototype human L-Cys/CSSC degrading enzyme* Cyst(e)inase) and found that bi-weekly administration to mice results in near complete depletion of the serum L-Cys/CSSC pool; **(ii)** mediates sustained and complete cessation of PCa tumor growth with cancer cells experiencing cell cycle arrest accompanied by increased oxidative stress and autophagy and **(iii)** importantly treatment for over a month was very well tolerated with no weight loss or gross toxicities.

Glycoengineered Outer Membrane Vesicles Displaying O-Polysaccharide Antigens Elicit Protection Against Francisella Tularensis

Chen, L., Cornell University
Baker, J., Cornell University
Huang, J. C. J., Amgen
Endicott, C., University of Connecticut
Putnam, D., Cornell University
Jones, B.,
DeLisa, M. P., Cornell University

The O-polysaccharide antigen (O-antigen) component of lipopolysaccharide on the surface of Gram-negative bacteria is both a virulence factor and a B-cell antigen. Antibodies elicited by O-antigens often confer protection against infection; hence, O-antigen conjugate vaccines have been designed against multiple pathogens. However, conventional methods for natural extraction or chemical synthesis of O-antigens are technically demanding, inefficient, and expensive. Furthermore, because carbohydrates are weak stimulators of the adaptive immune system, current carbohydrate-based vaccines require significant downstream processing to increase their immunogenicity. Here, we describe a scalable glycoengineering technique where recombinant O-antigen biosynthesis is coordinated with vesiculation in lab strains of *Escherichia coli* to produce acellular outer membrane vesicles (OMVs) decorated with pathogen-specific O-antigens on their surfaces. Because genes required for the biosynthesis of a given O-antigen are usually located in a distinct cluster, they can be easily amplified and transferred into *E. coli* for efficient, large-scale production. Moreover, the use of OMVs greatly simplifies glycoconjugate assembly and purification and results in a self-adjuvanting O-antigen delivery system. Immunization of mice with engineered OMVs displaying O-antigens from *Francisella tularensis* subsp. tularensis Schu S4 protected against lethal challenge with *F. tularensis* subsp. tularensis Schu S4, demonstrating the vaccination potential of our glycoconjugate vesicles.

Engineering and Preclinical Development of a Human Enzyme for Cancer Therapy Via the Depletion of the Serum L-Cysteine Pool

Stone, E.,
Cramer, S., The University of Texas at Austin
Saha, A., University of Texas at Austin
Tiziani, S., The University of Texas at Austin
Tadi, S., The University of Texas at Austin
Digiovanni, J.,
Georgiou, G., The University of Texas at Austin

Extensive studies have revealed that several high mortality cancers including prostate carcinomas (PCa), glioblastomas, and chronic lymphocytic leukemia (CLL) are critically dependent on the uptake of the non-essential amino acid L-cysteine (CYS) (and its oxidized form L-cystine (CSSC)) for proliferation, survival, and even resistance to many standard of care chemotherapeutics. The observation that relative to non-malignant tissues a tumor's excessive requirement for CYS/CSSC is a critical metabolic vulnerability, led us to exploit this tumor deficit by developing a novel therapeutic that systemically depletes CYS and CSSC. An important point is that most if not all normal tissue can make CYS from the essential amino acid L-Methionine, thus tumors are selectively impacted when CYS is unavailable.

We engineered and developed a human enzyme with the requisite kinetic and pharmacological properties that can (**i**) deplete the serum pool of CYS/CSSC to near completion, (**ii**) selectively and potently kill cancer cells through increased ROS production and (**iii**) mediate a sustained cessation of tumor growth without any weight loss or apparent toxicities.

The outstanding results from this approach have accelerated our efforts towards bringing this therapeutic to the clinic. We discuss the many studies currently ongoing and highlight the engineering, pharmacological studies, efficacy studies, biological mechanisms and IND enabling work in mice and non-human primates. Having a pharmacological tool to non-toxically manipulate the major extracellular redox buffer (CYS/CSSC) in living mammals and revoke a tumors ability to cope with its own reactive oxygen species is an unprecedented advance for oncology.

Towards Biofuel Production in Synechocystis Sp. PCC 6803: Expanding the Molecular Biology Toolbox for Pathway Engineering

Peebles, C. A. M., Colorodo State University

Biofuel production in microorganisms has great potential to fulfill this nation's need for fungible "green" transportation fuels. Because of the simplicity of bacterial cells, photosynthetic cyanobacteria make good candidates as a platform for biofuel production. *Synechocystis* sp. PCC 6803 is an attractive production platform due to rapid doubling times, a sequenced genome and an established molecular biology toolbox. This molecular biology toolbox was established to probe the biology of photosynthesis and has not been optimized for pathway engineering which is necessary for biofuels production. This presentation will discuss our efforts to develop an alternative counter selection marker and to develop and characterize promoters for gene expression under varying conditions. The use of these tools to study and optimize free fatty acid production in *Synechocystis* will also be discussed.

Deciphering Transcriptional Regulation Patterns for Novel Enzyme Discovery

Solomon, K. V., University of California, Santa Barbara
Haitjema, C., University of California, Santa Barbara
Henske, J. K., University of California, Santa Barbara
Borges-Rivera, D., Broad Institute of Massachusetts Institute of Technology and Harvard
Thompson, D. A., Broad Institute of Massachusetts Institute of Technology and Harvard
O'Malley, M. A., University of California, Santa Barbara

Anaerobic gut fungi are attractive microbes to engineer for bio-based chemical production from lignocellulose, and are an untapped reservoir for the discovery of new biomass-degrading enzymes. These fibrolytic microbes secrete an array of powerful cellulases and cellulolytic complexes (cellulosomes) for synergistic hydrolysis of plant biomass in herbivores, yet little sequence information exists for gut fungi. Though fungal hydrolytic activity has been shown to be substrate dependent, the underlying regulation mechanisms that coordinate the action of cellulases and cellulosomes from gut fungi remain unknown. We hypothesize that cellulose-degrading machinery is catabolite-repressed to conserve cellular energy, and our objective is to exploit this regulation mechanism to discover novel enzymes. To address this hypothesis, we have combined next-generation sequencing and proteomic approaches to examine cellulose-degrading enzyme production in a panel of gut fungi isolated from natural ecosystems. All fungi exhibited high enzymatic reactivity against a range of cellulosic and lignocellulosic substrates (filter paper, Avicel, reed canary grass), which was repressed in the presence of simple sugars. Through strand-specific RNAseq and use of the TRINITY assembly platform, we were able to assemble thousands of novel genes *de novo* from >27,000 transcripts without the need for genomic sequence information. RNA-Seq also elucidated global regulatory patterns in response to catabolite repression of biomass degradation, and in response to growth on cellulosic substrates of increasing complexity. Through these efforts, we have identified hundreds of transcripts encoding novel enzymes for biomass degradation, and the fungal transcriptome is particularly rich in GH6, GH48, and GH43-containing enzyme domains. As hypothesized, most of these transcripts are strongly repressed by the addition of simple sugars and are clustered within distinct ‘regulons’ of coordinated gene expression. Gene set enrichment analysis confirms the upregulation of these regulons across cellulosic substrates, and complementary proteomic analyses reveals that the composition the fungal cellulosomes is tuned by the presence of different substrates. More importantly, the functional enrichment of these regulons suggests a critical role for the divergent and unannotated transcripts that they contain. A dozen co-regulated transcripts from our screen bear no homology to known enzymes, and likely harbor previously undiscovered glycosyl hydrolase domains from nature, which we are currently investigating through protein crystallography. Collectively, this information will establish the molecular framework for anaerobic fungal hydrolysis, ultimately allowing us to refactor this system in anaerobic fungi and beyond.

Design and Engineer L-Amino Acid Deaminase from Proteus vulgaris As a Robust Biocatalyst for One-Step Biosynthesis of a-Keto Acids

Liu, L., Jiangnan University

Design and engineer L-amino acid deaminase from *Proteus vulgaris* as a robust biocatalyst for one-step biosynthesis of a-keto acids

Long Liu

School of Biotechnology,Jiangnan University, Lihu road 1800, Wuxi, China

Tel.: +86-510-85918312, Fax:+86-510-85918309,E-mail: longliu@jiangnan.edu.cn

Abstract

L-Amino acid oxidases (LAAOs), which catalyze the stereospecific oxidative deamination of L-amino acids to Î±-keto acids and ammonia, are flavin adenine dinucleotideâ??containing homodimeric proteins [1, 2]. L-Amino acid oxidases are widely distributed in diverse organisms and have a range of properties. In this work, we systematially engineered the LAAO from *Proteus vulgaris* as a robust biocatalyst for the efficient one-step biosynthesis of a-keto acids, including Î±-ketoglutaric acid, Î±-keto-Î³-methylthiobutyric acid, and phenylpyruvic acid, which are currently produced by multi-step chemical synthesis with heavy environmental pollution. First, the directed evolution of LAAO was done by error-prone polymerase chain reaction (PCR), and then the site-saturation mutagensis was conducted at the mutation sites of the positive mutants yielded by error-prone PCR. The biotransformaton conditions including biocatalyst concentration, the substrates concentration, temperature, and pH and so on. The maximum titers of Î±-ketoglutaric acid, Î±-keto-Î³-methylthiobutyric acid, and phenylpyruvic acid are 12.3 g/L, 63.6 g/L, and 11.8 g/L, respectively, under the specific optimal conditions. Compared with the traditional multi-step chemical synthesis, our one-step biocatalytic production of a-keto acids has an advantage in terms of less environmental pollution and thus has great potential for industrial production of a-keto acids.

[1]Liu L et al., Appl Microbiol Biotechnol. doi: 10.1007/s00253-013-5444-2. 2014.

[2]Hossain GS et al., J Biotechnol. 169, 112-120, 2014.

Key-words: L-amino acid deaminase, *Proteus vulgaris*, a-keto acids, protein engineering

Reducing Uncertainty in Metabolic Networks through Thermodynamics Based Flux Analysis

Kiparissides, A., Imperial College
Hatzimanikatis, V., Swiss Federal Institute of Technology of Lausanne (EPFL)

Almost every process or action that occurs within a cell involves a metabolic reaction. Knowledge of the metabolic state of a cell and how it responds to various stimuli and extracellular conditions can offer significant insight in the regulatory functions and how to manipulate them. The increasing availability of large metabolomics datasets enhances the need for computational methodologies that can organize the data in a way that can lead to the inference of meaningful relationships. Metabolic models comprising the entirety of reactions known within a pathway and/or cell provide an increasingly popular and effective platform to study the internal states of a cell. However unless all intracellular fluxes are known, constraint-based models, by design, lead to underdetermined problem formulations. This results in the existence of numerous alternate internal flux distributions that satisfy the mass balance constraints and can achieve the same optimum. Experimental information, when available, allows for some of these solutions to be discarded. Reaction fluxes that are not measured experimentally are allowed to vary between physiologically relevant bounds often derived from literature. The size of the resulting solution space, comprising the gamut of alternate optimal flux distributions, can be viewed as a representation of the uncertainty in predicting an exact intracellular state relevant to the studied physiology. The space of steady state flux solutions under uncertainty has attracted significant scientific interest in a number of studies that either attempted to correlate biological significance with characteristics of the solution space or attempted to achieve more accurate flux distributions by incorporating additional physicochemical information such as thermodynamic and molecular crowding constraints. However the problem of identifying network components with a significant contribution towards the observed uncertainty remains far from trivial.

In the present work we combine thermodynamics based constraint based modeling, Monte Carlo sampling, Design of Experiments (DoE) and global sensitivity analysis (GSA) in order to rank metabolites comprising a metabolic network based on their ability to constrain the gamut of possible solutions to a limited, thermodynamically consistent set of internal states. The proposed approach is modular and can be applied to a single reaction, a metabolic pathway or an entire metabolic network. This is, to our knowledge, the first attempt to use metabolic modeling in order to provide a significance ranking of metabolites to guide experimental measurements.

Controlling Local Substrate Concentrations and Enzyme Kinetics through Rationally Designed Intermolecular Interactions

Gao, Y., UC Riverside
Zhu, J., UC Riverside
Lin, J. L., UC Riverside
Wheeldon, I., UC Riverside

In nature we find many examples of enzymes and multienzyme stuctures where catalysis is enhanced by well-designed molecular interactions between the enzymes and their substrates. Two compelling examples are the enzyme superoxide dismutase (SOD) and the bifunctional enzyme thymidylate synthase-dihydrofolate reductase (TS-DHFR). SOD, one of the fastest known enzymes ($k_{cat} = 1.5 \times 10^9\ M^{-1}s^{-1}$), uses charge complementarity to produce substrate-enzyme interactions that enhance enzyme kinetics by directing the substrate to the enzyme's active site. A positively charge patch on the surface TS-DHFR restricts diffusion of a negatively charged reaction intermediate to a pre-defined channel between two active sites, sequestering the intermediate along the enzyme's surface and preventing diffusion to the bulk. From a catalysis perspective these two examples suggest a new and potentially powerful strategy for enzyme engineering, the rational design of intermolecular interaction outside of an enzyme's active site that enhance and control enzyme kinetics.

In this work we engineer new enzyme structures with quantifiable binding interactions between the enzyme and its substrate. We hypothesize that the engineered molecular interactions will lead to increases in local substrate concentrations thereby enhancing enzyme catalysis. We confirm this hypothesis by demonstrating control over the apparent Michaelis constant (K_M) of horseradish peroxidase (HRP) modified with a double stranded DNA structure that exhibits sequence dependent binding of phenolic HRP substrates. We extend this work to a second experimental system and demonstrate enhanced co-factor binding with an alcohol dehydrogenase (AdhD) through rationally designed molecular interactions between the NAD^+ mimic nicotinamide mononucleotide (NMN^+) and a double stranded DNA structure conjugated near the enzyme's active site. Using this system, we also vary the location of the DNA conjugation site to explore the effects of DNA location of kinetic enhancements and increased local substrate concentrations. In the case of HRP we achieve a nearly 3-fold decrease in $K_{M,app}$ resulting in a similar increase in enzyme efficiency (k_{cat}/K_M). In the case of AdhD we were able to reduce co-factor binding by nearly one half. These findings represent an important first step towards developing a new strategy of enzyme engineering by means of controlled substrate-enzyme interactions.

Discovery of a Pre-Eclampsia Associated Antibody That Binds to a Viral Antigen and Human Protein

Elliott, S. E., University of California, Santa Barbara
Soffici, A. R., Cottage Health System
Daugherty, P. S., University of California, Santa Barbara

Diverse diseases, including autoimmune conditions such as celiac disease and myasthenia gravis [1], have been proposed to involve environmental factors. While the immune system archives the response to environmental exposures in the form of an antibody repertoire, it has been difficult to impartially identify which molecules from this repertoire are associated with disease. To address this issue, we screened bacterial-displayed peptide libraries to identify peptides with enhanced binding to antibodies present in a diseased state over normally present antibodies and characterized the disease-associated binding motifs. We applied this quantitative discovery process to identify molecular diagnostics for pre-eclampsia (PE), a condition with unknown etiology that affects 5-8% of pregnancies. Previously characterized autoantibodies that bind the angiotensin II type 1 receptor in PE patients [2][3] are difficult to detect, vary in prevalence amongst studies, and most importantly, lack specificity. Thus, we hypothesized that additional PE-specific antibody biomarkers may exist.

Screening a fully randomized peptide library against pools of diluted plasma from PE and healthy outcome pregnancies (HOP) reproducibly identified a peptide binding motif with strong similarity to a region of a common viral antigen (VA), and we used directed evolution to expand this motif. Additionally, by combining peptide library screening with next-generation sequencing, we profiled antibody binding specificities of individuals and confirmed enrichment of this viral antigen linked motif amongst PE patients. Importantly, we determined that antibody binding to a bacterial-displayed 15-mer fragment of the VA containing this motif (VA15) correctly distinguished 67% of PE patients (n=42) and 86% of HOP (n = 43) at or near term. Furthermore, we linked this motif and verified antibody binding activity to a fragment of a human G protein-coupled receptor (hGPCRa), while the synthetic VA15 competed with this antibody binding. Moreover, VA15 inhibited antibody binding to the full-length hGPCRa overexpressed in HEK293T cells and natively expressed in a model trophoblast cell line. Finally, we verified the presence of hGPCRa in PE placental tissue.

Our results demonstrate that an unbiased quantitative molecular separation process identifies disease-specific antibody binding peptide motifs. Furthermore, these binding motifs can be used to identify environmental antigens and human protein targets of antibodies associated with a disease. Thus, these antibody detecting peptides can serve as novel molecular diagnostic tools and help elucidate potential therapeutic targets.

[1] Cusick, M., Libbey, J. and Fujinami, R. Molecular Mimicry as a Mechanism of Autoimmune Disease. *Clinical Reviews in Allergy and Immunology* **42,**102–111 (2012).

[2] Wallukat, G. *et al.* Patients with Preeclampsia Develop Agonistic Autoantibodies against the Angiotensin AT1 Receptor. *The Journal of Clinical Investigation* **103.**945–

952 (1999).

[3] Zhou, C. C. *et al.* Angiotensin Receptor Agonistic Autoantibodies Induce Pre-eclampsia in Pregnant Mice. *Nature Medicine* **14,** 855–862 (2008).

Study of the Effects of Alternative Fluxes on the Metabolic Control Analysis of Optimally Grown E. coli

Fengos, G., Swiss Institute of Bioinformatics
Ataman, M., Ecole Polytechnique Federale de Lausanne (EPFL)
Hernandez Gardiol, D., EPFL
Miskovic, L., Ecole Polytechnique Fédérale de Lausanne (EPFL)
Hatzimanikatis, V., Swiss Federal Institute of Technology of Lausanne (EPFL)

The intracellular metabolite and flux states have a significant impact on the responses to metabolic networks to changes in their enzyme activities and environmental conditions. Therefore it is important to characterize systematically all possible intracellular flux states, and analyse their properties. We have developed a novel framework called Flux Directionality Profile Analysis (FDPA), which enumerates all the possible intracellular flux profiles that are consistent with an observable set of extracellular fluxes. In the current study, we applied FDPA and we found the alternative intracellular flux profiles in *E. coli* under optimal aerobic growth condition. We next applied the ORACLE (Optimization and Risk Analysis of Complex Living Entities) framework to perform metabolic control analysis and identify the differences in the rate limiting reactions between the different intracellular flux profiles. We will discuss how the assumptions about alternative intercellular states can lead to distinct and different conclusions about the metabolic design strategies.

Intracellular Trafficking and Degradation of Antibody Drug Conjugates

Maass, K., Massachusetts Institute of Technology
Wittrup, K. D., Massachusetts Institute of Technology

Intracellular Trafficking and Degradation of Antibody Drug Conjugates

Katie Maass, K. Dane Wittrup

Chemotherapeutics used for cancer treatment are commonly limited in dose due to systemic toxicity. A new class of cancer therapeutics known as Antibody Drug Conjugates (ADCs) has shown promise in effectively killing cancer cells while reducing systemic toxicity. ADCs are comprised of a small-molecule cytotoxic drug conjugated to an antibody that targets an antigen specific to cancer cells. Ideally, the cytotoxic drug remains conjugated to the antibody until it is internalized within a cancer cell and then released to have its cytotoxic effect. However, there are numerous steps before the active drug component reaches its target in the cell, and the impact of ADC design parameters on these processing steps is poorly characterized. Increased understanding of their cellular processing may facilitate more rational design of ADCs.

In this work, we develop a kinetic model that describes the internalization and degradation of ADCs on a cellular level. Using flow cytometry and a novel gel imaging assay, we measure surface association and disassociation, internalization, and intracellular degradation and exocytosis rates for model fluorophore-labeled ADCs. Thus, we present a generalizable method to determine kinetic parameters, which can be used in a basic model for cellular processing of ADCs and can be incorporated into larger scale pharmacokinetic/pharmacodynamic models.

Ligand Responsive Hybrid Promoters and Bidirectional Promoters for Metabolic Engineering in Oleaginous Yeast Yarrowia Lipolytica

Shabbir-Hussain, M., Clemson University
Gambill, L., Clemson University
Williams, S., Clemson University
Blenner, M. A., Harvard Medical School & Children's Hospital Boston

Yarrowia lipolytica is a non-conventional, oleaginous yeast that grows on numerous substrates, including glycerol, fats, and other hydrophobic carbon sources. As an efficient producer of citric acid, lipase, and triacylglycerides, *Y. lipolytica* has gained attention as a bio-production platform. While genetic tools for constitutive Y. lipolytica promoters are available, fine tuning metabolic engineering of *Y. lipolytica* would benefit greatly from inducible promoters of defined strength. As an oleaginous yeast, engineering fatty acid synthesis and metabolism is of great interest, and there is considerably little information on the molecular details of fatty acid regulated gene expression. Here, we describe our work create fatty acid responsive hybrid promoters through two means. We first truncated endogenous beta-oxidation promoters that are known to be responsive to fatty acids. Through a series of promoter truncations of two fatty acid responsive genes, acyl-CoA oxidase 2 (*POX2*), and acetyl-CoA acetyl transferase (*PAT1*), we identified key elements in fatty acid regulated promoters including upstream activating sequences (UAS) and core promoter elements that confer fatty acid responsiveness. GFP reporter expression and quantitative PCR measurements were made to characterize promoter activity. The second method used employs the UAS sequences from the XPR2 gene. We describe how tandem repeats of these UAS sequences act as a fatty acid responsive promoter when combined with transcriptional start site. We subsequently leveraged the bidirectional nature of these UAS sequences to create a more compact dual-expression vector. This dual-expression system enables more genes to be expressed in Y. lipolytica with smaller plasmids. With new well-defined inducible promoters and expression systems, metabolic engineering of fatty acid regulated pathways will increase yield, productivity, and the scope of molecules that can be produced by *Y. lipolytica*.

Development of SNAP-Tagged Antibodies for Magnetic Bead-Based Immuno-PCR

Gollihar, J.,
Pothukuchy, A., University of Texas at Austin
Byrom, M.,
Ellington, A., University of Texas at Austin

Recombinant antibodies are frequently joined to other molecules or to surfaces via conjugation. Many conjugation methods rely on random modification of sulfhydryl, amine, or carboxyl groups on the antibody surface, but these modifications can affect the stability, solubility, and affinity of the antibody.

SNAP-tag technology developed by the New England Biolabs (Ipswich, MA), allows the human DNA-repair enzyme O(6)-alkylguanine DNA alkyltransferase (AGT) to be fused to proteins and to capture para-substituted O(6)-benzylguanine (BG) derivatives through a nucleophilic substitution reaction that releases free guanine. By coupling AGT to antibodies it should prove possible to homogeneously label the antibodies. This can in turn assist in developing sensitive, reproducible and accurate diagnostic assay methods.

We have successfully designed and developed SNAP-tag fusion antibodies against several cancer biomarkers such as Tumor Necrosis Factor (TNF), Granulocyte Macrophage Colony Stimulating Factor (GMCSF), and Interferon Gamma (IFNg). Using BG functionalized beads and oligonucleotides, we have produced covalently-modified capture and detection antibodies. In a bead-based sandwich immuno-PCR assay with optimized qPCR conditions we have detected as little as 35 attomoles of a protein target. We are currently optimizing a panel of oligo-conjugated or aptamer-conjugated antibodies and magnetic bead conjugates to develop a generalizable method for immuno-PCR reagent production.

Pigment-Based, Low-Cost, Portable Micronutrient Status Tests Using Engineered Bacteria

Watstein, D., Georgia Institute of Technology
Styczynski, M. P., Georgia Institute of Technology

Micronutrient deficiencies are a significant healthcare concern across the globe. Significant even in some developed nations, micronutrient deficiencies are more severe in the developing world and locally in the wake of major disasters. These conditions, though easily treated, remain a problem because they are often difficult to recognize and diagnose, requiring lab tests that are prohibitively expensive in both material and human resources for those in developing or remote areas.

As obligate consumers of the same micronutrients, bacteria possess cellular machinery to control intracellular micronutrient levels and have corresponding regulatory mechanisms to respond to varying concentrations in their environment. We have developed a whole-cell bacterial biosensor harnessing these properties for use in a diagnostic test for blood micronutrient status, specifically for the detection of zinc levels in blood. The test is designed to be inexpensive in total cost of operation, requiring no complex equipment and minimal medical training to administer and interpret. This would obviate the logistical problem of laboratory access and sample transport in remote and low-resource environments, allowing on-site diagnosis of micronutrient deficiencies in the populations most at risk.

To create this biosensor, we designed and implemented genetic circuitry to trigger specific changes in color that are visible to the naked eye in response to zinc levels. In one implementation, we have used pigments as readouts for the biosensor. With appropriate transcriptional control, pigment-based color changes could be essentially switch-like, reducing ambiguity in the interpretation of test results. Moreover, production of even a small amount of pigment-producing enzyme could enable synthesis of sufficient pigment to be visible to the naked eye, potentially decreasing the total time to coloration and the biochemical burden on the cell. In another implementation, we have used chromoproteins as indicators. Chromoproteins have different potential advantages, including that they require substantially less DNA, which simplifies cloning and other molecular biology manipulations. Here we present results from our bacteria that produce multiple visible colors in response to zinc levels, as well as the results of our efforts to tune these sensors using a variety of metabolic engineering strategies, including changes in promoters, ribosomal binding sites, and transporters.

High-Throughput Design of Regulatory Protein-Based Biosensors for Screening Biosynthesis Libraries

Cirino, P. C., University of Houston
Frei, C. S., University of Houston
Qian, S., University of Houston

In vivo high-throughput screening systems for detecting select metabolites are useful for engineering enzymes and flux through biosynthesis pathways. Transcriptional regulator proteins serve as gene switches in microorganisms and can be applied as highly specific and sensitive endogenous molecular biosensors. For cases where there is no known regulator that responds to a metabolite of interest, protein engineering may be used to alter inducer specificity of an existing biosensor system. We combine ligand binding pocket saturation mutagenesis with Fluorescence Activated Cell Sorting based on fluorescent reporter protein expression to generate novel molecular biosensors based on the AraC regulatory protein from E. coli. We have isolated AraC variants with altered specificity toward a variety of small molecules such as mevalonate, p-coumaric acid, triacetic acid lactone (TAL), and other 2-pyrones. These novel biosensors and their use in screening for improved biosynthesis of the corresponding effector will be described.

Through the course of biosensor engineering we have optimized parameters to improve the design process and streamline high-throughput screening efforts. Examples include frequency of positive vs. negative sorts, time to sort induced cells, method of recovering post-sort clones, and half-life of the reporter protein. Next-Generation-Sequencing (NGS) was also used to monitor AraC library sequence evolution toward responsiveness to various ligands. Key results from these studies will be presented. Finally, a recently solved crystal structure of one AraC ligand binding domain variant (responding to TAL) reveals determinants of molecular recognition, and this will be discussed.

Intracellular FRET-Based Assay for Redesigning the Specificity of Secreted Proteases

Guerrero, J. L., University of California, Santa Barbara
O'Malley, M. A., University of California, Santa Barbara
Daugherty, P. S., University of California, Santa Barbara

Proteases constitute 2% of the human genome and regulate many biological processes including cell growth and migration, blood coagulation, and programmed cell death. A generally applicable, high-throughput strategy to engineer proteases to cleave a target substrate with high specificity and high catalytic efficiency would greatly expand the use of proteases for biotechnological and therapeutic applications. We have developed a cell-based assay for redesigning protease selectivity by screening protease mutant libraries for cleavage of a fluorogenic peptide substrate exhibiting Förster resonance energy transfer (FRET). Our screening method relies on using fluorescence activated cell sorting (FACS) for detecting cleavage of the FRET reporter substrate, which will allow screening of large protease mutant libraries (~10^7members) to identify the activity of interest. Here we report the application of this method to the protease human kallikrein 7 (hK7) to identify variants that selectively cleave the central hydrophobic core of the amyloid beta (Aβ) peptide, involved in Alzheimer's disease pathology.

Using flow cytometry, we optimized expression conditions to detect hK7 activity in yeast with a co-expressed intracellular Aβ8 (KLVF↓F↓AED) FRET substrate. hK7 displays modest activity towards the target Aβ8 substrate but prefers tyrosine (Y) at the P1 position. We hypothesized that amino acid substitutions within hK7 may yield variants that prefer the phenylalanine (F) at P1 of Aβ8 and exclude tyrosine, thereby narrowing the specificity towards Aβ8. We randomly mutated the hK7 gene using error-prone PCR to generate a library with an average of 2 base substitutions per gene. Screening this library for cleavage of the Aβ8 FRET reporter using fluorescence activated cell sorting yielded hK7 mutants with improved activity against the therapeutically relevant Aβ8 substrate. These mutants were then recombined using the staggered extension process (StEP) and the resulting library was screened to produce mutants with additive improvements in activity and specificity for the Aβ8 substrate. This methodology can be further applied to engineer other human proteases for highly specific degradation of proteins implicated in a disease state.

Talens Assisted Multiplex Engineering for Accelerated Genome Evolution in Saccharomyces Cerevisiae

Zhang, G.,
Wang, Q., Key Laboratory of Systems Microbial Biotechnology, Tianjin Institute of Industrial Biotechnology, Chinese Academy of Sciences

Genome engineering is an essential tool of constructing a genotype to achieve a desired phenotype, a challenge whose difficulty is influenced by the selectivity and scale of genomic alteration required. Here we report TALENs Assisted Multiplex Engineering (TAME) as a generally applicable genome engineering method for editing the genome and accelerating the genome evolution in yeast. Through the iterative cycles of TAME induced mutants coupled with efficient screening by fluorescence reporter, TAME can continuously and rapidly improve simple target trait by accumulating multiple beneficial genetic modifications in evolving yeast genome. In order to assess its efficiency further, we then show the application of TAME for improved ethanol tolerance, a key and complicated trait for microbial production of fuels. Five rounds of iterative TAME screening led to acting synergistically to confer the engineered yeast strain with substantially improved ethanol tolerance. Therefore, TAME is a broadly applicable platform for accelerated evolution for desired traits in eukaryotes.

A Novel, in Vivo Continuous Evolution Strategy in Saccharomyces Cerevisiae to Create Pathway and Protein Variants

Sun, J., University of Illinois at Urbana-Champaign
Crook, N., The University of Texas at Austin
Abatemarco, J., The University of Texas at Austin
Alper, H., The University of Texas at Austin

Current evolution techniques in yeast usually require *in vitro* synthesis of a mutant library, transformation of the created library into yeast cells, and selection of desired phenotypes towards a preferred pressure. Due to the difficulties of library synthesis and low efficiency of yeast transformation, the traditional directed evolution process is time-consuming and labor-intensive. Here, we developed a novel *in vivo* continuous evolution process in which mutagenesis occurs simultaneously with cell growth, resulting in a more streamlined evolution cycle. As a proof-of-concept, we applied *in vivo* continuous evolution to the development of improved variants of model proteins, global transcriptional regulators, and multigene catabolic pathways in *S. cerevisiae* over several rounds of *in vivo* optimization. Taken together, these demonstrations show that *in vivo* continuous evolution can significantly accelerate the pace of pathway development in yeast.

Generation of Protease-Inhibiting Monoclonal Antibodies By Novel Paratope Design and Function-Based Screening

Nam, D. H.,
Fang, K.,
Rodriguez, C., University of California, Riverside
Lopez, T., University of California, Riverside
Ge, X., University of California, Riverside

Generation of protease-inhibiting monoclonal antibodies

by novel paratope design and function-based screening

Dong Hyun Nam, Tyler Lopez, Kuili Fang, Xin Ge* Chemical and Environmental Engineering, University of California, Riverside 900 University Ave, Riverside, CA 92521

As extremely important signaling molecules, proteases precisely control a wide variety of physiological processes, and thus many diseases are associated with altered protease expression or substrate proteolysis. Consequently, proteases are one of the largest families of pharmaceutical targets, and an apparent pharmaceutical mechanism is to block abnormal or pathogenic proteolysis by inhibiting the catalytic reactions. Most therapeutic protease inhibitors currently in clinical use or under developments are peptides or their chemical compound mimics. Because ~2% of the human genome is estimated to encode proteases with at least 500-600 proteases have been identified, specificity is a crucial property for any protease inhibition therapy. However, achieving target specificity has been recognized as the main hurdle for protease inhibitor developments. This is because the reaction mechanism is highly conserved among the same class or family of proteases, resulting in lead compounds often inhibit multiple targets and potentially causing unwanted side effects. With these aspects, antibody-based inhibitors are emerging as very attractive therapeutic agents due to their exclusive specificity given by a large antigen-antibody contact surface.

Although isolating antigen-specific mAbs has become routine, developing antibodies with protease inhibitory functions is still challenging, because of (1) low antigenicity of the enzymatic active sites, (2) lack of a function-based selection method. This study devolved a panel of novel techniques to address these obstacles and used matrix metalloproteinase (MMP)-14 as the target, which plays important role in tumor growth, invasion and neovascularization.

Inspired by the fact that a large proportion of antibodies isolated from camels and llamas can bind to active pockets and inhibit enzymatic reactions, we analyzed inhibitory antibody sequences and structures, and hypothesized that convex shaped paratopes given by ultra long CDRs are inhibition-prone. Large synthetic human antibody libraries (>10^9 variants) carrying extended CDR-H3 ranging from 23 to 27 aa were constructed. After phage panning on MMP-14, dozens of Fab clones exhibiting strong binding signals were isolated, with nM range of affinity for many of them. CDR-H3 sequences of 40 randomly picked binding clones showed dramatic patterns with central histidines flanked by arginine rich sequences. These residues presumably chelate the catalytic zinc and recognize negatively charged surface of the reaction pocket of MMP-14, respectively. More strikingly majority of these sequenced Fabs demonstrated inhibitory activity with various potencies. Regarding to antibody selection method, current technology is essential binding-based assays with little controls on epitope specificity or inhibitory functions. In this study, we developed a function-based high-throughput screening method by achieving epitope-specific FACS that simultaneously performed selection on target protease (MMP-14) and counter-selection on its inhibitor (TIMP-2), which were conjugated with different fluorophores. Able to enrich inhibitory clones from non-specific or non-inhibitory clones, this method was applied for affinity maturation of inhibitory mAbs. In addition, we also applied motif grafting to convert a low affinity peptide inhibitor into a high potent mAb with 2000 folds of improvements in inhibition potency from 150 ÂµM to 70 nM. These novel techniques significantly advanced our understanding of enzymatic inhibition mechanisms and greatly facilitated the discovery of highly selective and highly potent inhibitory antibodies, thus are expected to have a broad range of applications in pharmaceutical and biotechnological industries. Furthermore, because the vulnerable sites for drug binding of many viral coating proteins represent concave conformations, the synthetic antibody libraries with convex paratopes developed in this study also have potentials for the development of neutralizing antibodies.

Directed Protein Evolution for Specificity and Affinity Towards Chemically Modified RNAs Via Secm Ribosome Display

Baldridge, K., University of Texas at Austin
Bryant-Friedrich, A., University of Toledo
Contreras, L., University of Texas Austin

- Application of directed evolution approaches to develop in vivo probes targeting modified RNAs

Application of directed evolution approaches to develop *in vivo* probes targeting modified RNAs

Recent research has brought ribonucleic acids (RNAs) to the forefront of understanding how cellular processes are controlled with the ever-expanding list of non-coding RNAs (ncRNAs)[1â??4]. In the vast majority of cases, ncRNAs enact their functions by binding with DNA, proteins, and other RNAs to effect a change[3,5]. These interactions can be very sensitive to small changes in sequences or chemical modifications. Indeed, post-transcriptionally modified RNAs have long been known to be of essential importance in a variety of well-studied cellular RNAs including both tRNAs and rRNAs. Furthermore, more recent studies with the widely distributed 6-methyladenine (m_6A) base modification have shown their importance in the vast majority of cellular RNAs, including both coding and noncoding RNAs[6].

Until recently, techniques for studying modified RNAs have largely been limited to mass spectrometry for identification of novel modifications as well as enzyme knockouts for studying the function of known RNA modifications. Recently, high-throughput techniques such as bisulfite sequencing[7] and m_6A-seq[8] have advanced the field rapidly by allowing whole-genome profiling of RNA modification sites. However, these techniques analyze RNA samples after extraction outside of the in vivo environment and, as such, are susceptible to changing chemistry which can confound identification of modification types and locations[9,10]. Therefore, the ability to capture modified RNAs in the cellular context would reduce the likelihood of artificial chemical modifications by protecting RNAs during extraction while allowing similar sequencing approaches which are already in use for identification of novel modification sites.

In this work, we are taking advantage of directed evolution approaches to engineer natively expressing *Escherichia coli* proteins for high affinity and specificity against chemical modifications in RNAs. The scheme was tested using a well-studied protein, AlkB (alpha-ketoglutarate dependent dioxygenase), which naturally functions as a repair enzyme for m_3C and m_1A base methylations in DNA and RNA[11,12]. Initially, the display and selection system were validated using pre-defined pseudolibraries containing two previously characterized mutants of AlkB; alanine substitutions at amino acids 135 (D135A) and 69 (W69A) both destroy catalytic function while increasing or decreasing affinity for the modified RNA, respectively. Pools of these two displayed proteins in approximate ratios ranging from 1:100 to 1:10,000 were screened over nonspecifically methylated (methyl-methanesulfonate-treated) RNA targets to demonstrate enrichment of the higher affinity mutant D135A from a pool containing both mutants. Following validation of this approach, protein libraries were created from wild-type and mutant variants of AlkB using error-prone PCR. For directed evolution using full size libraries, synthetic 30-mer RNA containing m_1A at a specified nucleotide was used as the target for screening. Our initial results show that specific affinity for binding modified RNAs can be engineered into native *E. coli* proteins.

1. Storz, G., Vogel, J. & Wassarman, K. M. Regulation by small RNAs in bacteria: expanding frontiers. *Mol. Cell* 880â??91 (2011).
2. Salta, E. & De Strooper, B. Non-coding RNAs with essential roles in neurodegenerative disorders. *Lancet Neurol.* 189â??200 (2012).
3. Mercer, T., Dinger, M. & Mattick, J. Long non-coding RNAs: insights into functions. *Nat. Rev. Genet.* 155â??159 (2009).
4. Esteller, M. Non-coding RNAs in human disease. *Nat. Rev. Genet.* 861â??74 (2011).
5. Huang, Y. *et al.* Molecular functions of small regulatory noncoding RNA. *Biochem.* 221â??230 (2013).
6. Sibbritt, T., Patel, H. R. & Preiss, T. Mapping and significance of the mRNA methylome. *Wiley Interdiscip. Rev. RNA* 397â??422 (2013).
7. Schaefer, M., Pollex, T., Hanna, K. & Lyko, F. RNA cytosine methylation analysis by bisulfite sequencing. *Nucleic Acids Res.* e12 (2009).
8. Dominissini, D. *et al.* Topology of the human and mouse m6A RNA methylomes revealed by m6A-seq. *Nature* 201â??6 (2012).
9. Hofer, T. & MÃ¶ller, L. Optimization of the Workup Procedure for the Analysis of 8-Oxo-7,8-dihydro-2â??-deoxyguanosine with Electrochemical Detection. *Chem. Res. Toxicol.* 426â??432 (2002).
10. Warnecke, P. M. *et al.* Identification and resolution of artifacts in bisulfite sequencing. *Methods* 101â??107 (2002).
11. Van den Born, E. *et al.* Viral AlkB proteins repair RNA damage by oxidative demethylation. *Nucleic Acids Res.* 5451â??61 (2008).
12. Mishina, Y., Chen, L. X. & He, C. Preparation and characterization of the native iron(II)-containing DNA repair AlkB protein directly from Escherichia

coli. *J. Am. Chem. Soc.* 16930â??6 (2004).

Engineering Secretion Machinery for High-Throughput Protein Production

Tullman-Ercek, D., University of California

Protein production traditionally relies on cytosolic accumulation, such that extensive purification steps are required to separate the desired product from the intracellular milieu. During protein engineering experiments, it is often desirable to characterize many variants, and the limited throughput of production by cytosolic accumulation therefore limits the number of variants that can be examined. We envision using secretion to enable high-throughput protein production and analysis of libraries of proteins. Secretion offers the benefits of simplified purification steps, but low titers have limited the applicability of this strategy, particularly in bacterial hosts. To this end, we recently engineered a type III secretion system to achieve as much as a 100-fold increase in secretion titer of heterologous proteins, giving rise to titers of >100 mg/L of protein in the extracellular space. We applied the engineered system to the production of various classes of proteins, including enzymes, biomaterial-forming proteins [1], and antimicrobial peptides. After a single purification step, the secreted proteins were >90% pure as compared to 35-65% purity from cytosolic purification. Moreover, after secretion the enzymes were active and the biopolymer-forming proteins could be cross-linked to form materials with highly reproducible and tunable properties. We note that a few of the heterologous proteins tested were incompatible with secretion; we are currently establishing the key parameters that govern "secretability" via the type III machinery. In this presentation, I will describe the three-pronged strategy used to improve secretion titers: 1) gaining control over secretion machinery regulation [2], 2) engineering the needle-forming protein, and 3) adding components to the media. I will also discuss our development of a model that uses the primary sequence to predict which heterologous proteins will be secreted, and detail how we applied this enhanced system to develop antimicrobial hydrogels for tissue culture applications.

1. Azam A., Metcalf K..J., Li C., Tullman-Ercek D. "Type III Secretion as a Generalizable Strategy for the Development of Peptide-based Materials." (Submitted).

2. Metcalf K.J., Finnerty C., Azam A., Valdivia E., Tullman-Ercek D. (2014) "Using transcriptional control to increase titer of secreted heterologous proteins by the type III secretion system." Appl. Environ. Microbiol. (In press).

Genome-Wide Search for Intracellular Factors Affecting RNA Folding Via in Vivo Oligonucleotide Hybridization

Vasquez, K.,
Vazquez-Anderson, J., University of Texas at Austin
Hatridge, T.,
Contreras, L. M., The University of Texas at Austin

Specific intracellular factors (e.g. proteins, metal ions, etc.) help RNAs reach their functionally active structures. However, detecting these factors and unveiling their influence on folding is challenging, particularly *in vivo*. Hereby, we present a novel approach where a transposon library is screened to detect cellular factors influencing the model Tetrahymena group I (gI) intron folding *in vivo* in *Escherichia coli* by utilizing a fluorescence-based oligonucleotide hybridization assay. Our *in vivo* structural probing assay utilizes a riboregulator that expresses green fluorescence protein (GFP) when oligonucleotide hybridization occurs between the probe and a structurally accessible local region within a target RNA, providing valuable information on RNA folding. We have profiled the gI intron *in vivo* in both wild type *E. coli* cells and in an StpA mutant strain; the suppressor of *td* phenotype A (stpA) protein acts as an RNA folding chaperone for the gI intron. Specifically, when probing a central region (P3-P4) that is critical for folding of the intron, we have confirmed that this factor influences gI intron folding, as previously reported in vitro. This region undergoes transient reconfigurations giving rise to low abundance folding intermediates, one of which is termed alt-P3. We will discuss these findings as well as the application of this intracellular fluorescence-based system for a genome-wide search of other factors that affect gI intron folding *in vivo*. In particular, we probe the dynamic P3-P4 region of the gI intron in screening a transposon library. Significant differences in fluorescence relative to the wild type cell's fluorescence pattern can be accredited to inhibited or enhanced RNA folding, affected by specific cellular factors.

An Alternative Splicing Platform for Programming Protein Function

Mathur, M., Stanford University
Smolke, C. D., Stanford University

Our ability to effectively engineer biological systems is limited by the tools and strategies available to detect, transmit, and control molecular information. While a number of RNA-based tools have been developed for controlling gene expression, molecular platforms that support more sophisticated mechanisms of control, such as reprogramming protein form and function, are needed to advance the scale and complexity with which we can construct and interact with biological systems. Alternative splicing is a naturally occurring process in eukaryotes that generates multiple protein isoforms from a single coding region by altering the ways in which exons are joined together. Alternative splicing is a major source of protein diversity and greatly increases genomic coding capacity, such that over 95% of multiexon genes in humans are alternatively spliced. A platform for programming alternative splicing has the potential to broaden current regulatory capabilities by providing a strategy to dynamically program protein function. We have developed a design framework that supports the modular and extensible implementation of an alternative splicing regulatory platform. We have constructed and characterized an intron framework that produces defined mutually exclusive alternative splicing patterns when placed in the context of various exon sequences. By integrating ligand-responsive synthetic RNA switches into this intron framework we demonstrate the conditional regulation of alternative splicing patterns. This alternative splicing platform has been used to conditionally link functional domains in modular proteins, such as TALE-TFs, dynamically program protein function, and ultimately engineer complex regulatory networks in higher organisms.

Transformable Facultative Thermophile Geobacillus Stearothermophilus NUB3621 As a Host Strain for Metabolic Engineering

Blanchard, K., Emory University
Robic, S.,
Matsumura, I., Emory University School of Medicine

Protein engineering requires reliable expression of the protein by host cells. Engineering can be impeded by the misfolding of heterologously expressed proteins. *Geobacillus stearothermophilus* NUB3621 is a readily transformable facultative thermophile. It could be used to express and properly fold proteins derived from its many mesophilic or thermophilic Bacillaceae relatives or to direct the evolution of thermophilic variants of mesophilic proteins. Moreover, its capacity for high-temperature growth should accelerate chemical transformation rates in accordance with the Arrhenius equation and reduce the risks of microbial contamination. Its tendency to sporulate in response to nutrient depletion lowers the costs of storage and transportation. We have developed a draft genome sequence of *G. stearothermophilus* NUB3621 as well as inducible and constitutive expression plasmids that function in this organism. Utilizing these tools, we plan to evolve aminoglycoside phosphotransferase enzymes to be thermostable. We will then determine whether the thermostable variants are better able to adapt to a novel substrate, amikacin, than their mesophilic counterparts.

An Analytical Approach to Bistable Biological Circuit Discrimination Using Real Algebraic Geometry

Siegal-Gaskins, D., California Institute of Technology
Franco, E., Univeristy of California, Riverside
Zhou, T.,
Murray, R. M., California Institute of Technology

Small biomolecular circuits with two distinct and stable steady states have been identified as essential components in a wide range of biological networks, with a variety of mechanisms and topologies giving rise to their important bistable property. Understanding the differences between circuit implementations is an important question, particularly for the synthetic biologist faced with the challenge of determining which bistable circuit design out of many is best for their specific application. In this work we explore the applicability of Sturm's theorem—a tool from 19th-century real algebraic geometry—to comparing "functionally equivalent" bistable circuits without the need for numerical simulation. We consider two genetic toggle variants and two different positive feedback circuits, and show how specific topological properties present in each type of circuit can serve to increase the size of their operational range. The demonstrated predictive power and ease of use of Sturm's theorem suggests that algebraic geometric techniques may be underutilized in biomolecular circuit analysis.

Characteristics of Protein Incorporated within Poly(vinyl alcohol) Hydrogel Membrane

Salgado, A., Prairie View A&M University

A slow and controlled drug delivery system is vital to reducing the overall dosage amount and any toxic symptoms associated with the administration of drugs. The purpose of this research is to quantify and optimize the release of protein that has been incorporated within a Poly (vinyl alcohol) (PVA) membrane chemically cross-linked with Glutaraldehyde (GA) hydrogel to serve as a protein drug delivery system. A PVA-GA hydrogel membrane has a three-dimensional structure which allows the protein to permeate through the matrix. The open mesh created by GA results in advantageously high diffusivity that can be controlled by variations in the degree of crosslinking of the polymer matrix. As a part of a development of a drug release system, this study investigates the extent of protein release from the polymeric network of the hydrogel into buffer solution of varied pH. The studied parameters include: the concentration of cross-linking agent (glutaraldehyde) of 0.75 and 1% (v/v), the concentration of protein (0.4 and 0.7 mg/ml) and its overall effect on the swelling of the hydrogels, protein diffusion rate, and the release characteristics of the protein. Ovalbumin, a 45kD protein, was used as the model protein. The protein loaded membranes were characterized by Equilibrium Solution Content (ESC), Fourier Transform Infrared Spectroscopy (FTIR) and Thermal Gravimetric Analysis (TGA). The ESC experiments show that once protein is incorporated into the hydrogel matrix, the additional hydroxyl groups from the protein permits additional interaction with the buffer solution and caused an increase in swelling and the mesh size. A higher concentration of protein at 0.7mg/ml resulted in a varied mesh size value. This data suggests that the pores were overloaded and resulted in a dynamic environment which the protein moved in and out of the pores. Consequently, the 0.7mg/ml resulted in the highest release percentage of the diffusion experiments at 10% higher than the lower concentration of 0.4mg/ml. Also, the protein release demonstrates that from an initial pH of 7.4, ovalbumin prefers to diffuse into a buffer solution that is more alkaline of pH 8 than an acidic environment of pH 6.5. There was a 20% decline in the release within an acidic environment. The dissolution rate constant, k_H and diffusion coefficient, D, confirm that the higher pH is a preferred environment as reported. The FTIR show that the protein within the hydrogel cannot be visualized or detected once chemical crosslinking has occurred. However, the TGA results indicate that this is a temporary phenomenon because thermal properties of the hydrogel are not affected by the incorporation of protein but rather chemical crosslinking does affect the thermal properties. The results indicate that the protein was successfully incorporated and the protein does diffuse from the membrane over time. Furthermore, the interaction between the hydrogel and protein is temporary and confirms that the membrane system is ideal for protein drug delivery. Although drug release systems have been studied, our objective is to investigate the mechanism of release and determine the efficiency as it correlates to membrane structure and environment. In our future studies proteins of various sizes will be used to elucidate the protein release from the hydrogel network.

A Hybrid Riboswitch–Small RNA Molecule for Metabolic Engineering in an n-Butanol Pathway

Lahiry, A., The Ohio State University
Stimple, S. D., The Ohio State University
Wood, D. W., The Ohio State University
Lease, R. A., Ohio State University

A hybrid riboswitch - Small RNA molecule for metabolic engineering in an n-butanol pathway

Ashwin Lahiry[redacted][a], Samuel Stimple[b], David W. Wood[b] and Richard A. Lease[b]

(a) Department of Microbiology, (b) Department of Chemical and Biomolecular Engineering, The Ohio State University, Columbus, OH 43210, USA.

Biologically produced *n*-butanol is a realistic fuel alternative to gasoline[1] that can be utilized in existing infrastructure and vehicles. In fermentation using *Clostridium spp.*, balancing cell health with fermentation yield and *n*-butanol selectivity is a significant metabolic engineering problem. Gene knockouts to *ack* or *ptb* genes increase butanol yield by decreasing acetate or butyrate â??contaminants,â?□ but compromise cell health[1,2]. To address these challenges we are engineering bacterial small RNAs (sRNAs), flexible stress response regulators that rapidly fineâ??tune gene expression at low energetic cost to the cell[3]. These sRNAs specify their target mRNAs by complementary base pairing to decrease translation, thus decreasing target protein concentration and activity. In our work, we are utilizing the well-characterized *E. coli* sRNA DsrA as a platform for specific, targeted mRNA control, initially in *E. coli* but ultimately in *C. tyrobutyricum.* To assay DsrA sRNA function, we have created a plasmid-based *E. coli* system for inducible and scalable production of DsrA sRNA which natively regulates two mRNA targets (*rpoS* and *hns*) via distinct sRNA stem-loop structural domains.[4] The *rpoS* and *hns* mRNA leader sequences are fused to green and cyan fluorescent reporter proteins, respectively, and are also under independent, inducible, orthogonal and scalable promoter control. This system enables us to control and optimize desired sRNA levels to produce specific expression levels of two mRNAs, measured in real time in vivo, using a 96â??well microtiterâ??plate reader. We describe a quantitative responseâ??surface model for our system that demonstrates tunability of gene expression using a small RNA, and that can be used to predict desired sRNA regulatory output based on concentrations of individual gene-inducer ligands (IPTG, arabinose, and anhydrotetracycline). Further, we have used this system to analyze two types of hybrid riboswitch-sRNA variants which incorporate specific aptamer domains that bind ligands (theophylline or thiamine) based on their ability to activate or repress DsrA sRNA activity. Recent discovery of two natural RSâ??sRNAs [5,6] as well as previous work done in *E. coli* with engineered ncRNAs fused to RNA aptamers[7] confirm the utility of using a ligand/metabolite responsive sRNA in metabolic engineering applications. An advantage of using a RSâ??sRNA hybrid is that the pathway flux ultimately can be coupled to a ligand within the metabolic pathway that binds a riboswitch aptamer, for rapid metabolic-to-genetic feedback control.

References:

1. Zheng Y.N. *et al.* (2009). Problems with the microbial production of butanol. J Ind Microbiol Biotechnol 36. 1127-1138.
2. Yu M *et al.* (2011). Metabolic engineering of Clostridium tyrobutyricum for n-Butanol production. Metabolic engineering 13, 372-382.
3. Storz, G., Vogel, J. and Wassarman, K. (2011) Regulation by Small RNAs in Bacteria: Expanding Frontiers. Mol Cell 43:880-891.
4. Lease, R.A. and Belfort, M. (2000). A trans-acting RNA as a control switch in Escherichia coli: DsrA modulates function by forming alternative structures. Proc Natl Acad Sci U S A 97, 9919-24.
5. Sruti DebRoy et al. (2014) A riboswitch-containing sRNA controls gene expression by sequestration of a response regulator. Science 345:6199, 937-939
6. Mellin, J.R et al. (2014) Sequestration of a two-component response regulator by a riboswitch-regulated noncoding RNA. . Science 345:6199, 939-941
7. Lei Qi et al. (2012) Engineering naturally occurring trans-acting non-coding RNAs to sense molecular signals. Nucleic Acids Research, 1-12.

Bacterial Inner Membrane Display for Isolating Intracellular Antibodies from a Naïve Library

Moghaddam-Taaheri, P.,
Karlsson, A. J., University of Maryland

Antibodies that function intracellularly (intracellular antibodies) are promising tools for knocking down intracellular protein activity, but they present a protein engineering challenge because the reducing environment in the cytosol prevents formation of the disulfide bonds normally required for antibody folding and function. In our previous work, we developed an *Escherichia coli* inner membrane display platform that displays proteins that correctly fold in the *E. coli* cytoplasm by harnessing the intracellular folding quality control of the twin-arginine translocation pathway. Our present work used this platform to display a naïve library of single-chain variable fragment (scFv) antibodies with diversity in the third complementarity-determining regions (CDR3s) of the light and heavy chains. We then screened the library for binding to survivin, an apoptosis inhibitor that is upregulated in cancer cells. Our first round of screening isolated an scFv that binds to survivin and is expressed solubly in the *E. coli* cytoplasm. Using this isolate as the parent for random mutagenesis, we synthesized a second generation library and isolated additional scFvs that demonstrate improved binding to survivin and improved solubility compared to the intracellular scFv isolated in the first round of screening. Our results show the utility of the inner membrane display platform for isolating high-affinity, highly soluble intracellular antibodies that bind to clinically relevant targets.

Ensemble Modeling Identifies the Mechanism By Which Clpp Impacts NO• Defense Systems in E. coli

Brynildsen, M. P., Princeton University
Robinson, J. L., Princeton University

Nitric oxide (NO•) is an antimicrobial used by immune cells to neutralize pathogens. The importance of NO• to immune function is evidenced by the many pathogens, including *Mycobacterium tuberculosis*, *Neisseria meningitides*, *Vibrio cholerae*, *Salmonella enterica*, and enterohemorrhagic *Escherichia coli* (EHEC), that depend on NO• detoxification to establish an infection[1-7]. Inhibitors of NO• defense systems have the potential to form a novel class of anti-infectives; however, known agents are either toxic to humans or poorly transported into bacterial cells. Here I will discuss our discovery of ClpP as major mediator of NO• stress in *E. coli*, and how we used a quantitative model of NO• stress[8,9] within an innovative ensemble approach to elucidate the underlying mechanism. These results demonstrate the utility of quantitative modeling to the study of complex, systems-level stresses, such as NO•, and identify a novel target that when inhibited enhances the antimicrobial effects of NO• on *E. coli*.

References

1 Kakishima, K., Shiratsuchi, A., Taoka, A., Nakanishi, Y. & Fukumori, Y. (2007), Participation of nitric oxide reductase in survival of Pseudomonas aeruginosa in LPS-activated macrophages, *Biochem Biophys Res Commun*, **355**, 587-591.

2 Darwin, K. H., Ehrt, S., Gutierrez-Ramos, J. C., Weich, N. & Nathan, C. F. (2003), The proteasome of Mycobacterium tuberculosis is required for resistance to nitric oxide, *Science*, **302**, 1963-1966.

3 Kulasekara, B. R., Jacobs, M., Zhou, Y., Wu, Z., Sims, E., Saenphimmachak, C., Rohmer, L., Ritchie, J. M., Radey, M., McKevitt, M., Freeman, T. L., Hayden, H., Haugen, E., Gillett, W., Fong, C., Chang, J., Beskhlebnaya, V., Waldor, M. K., Samadpour, M., Whittam, T. S., Kaul, R., Brittnacher, M. & Miller, S. I. (2009), Analysis of the genome of the Escherichia coli O157:H7 2006 spinach-associated outbreak isolate indicates candidate genes that may enhance virulence, *Infect Immun*, **77**, 3713-3721.

4 Shimizu, T., Tsutsuki, H., Matsumoto, A., Nakaya, H. & Noda, M. (2012), The nitric oxide reductase of enterohaemorrhagic Escherichia coli plays an important role for the survival within macrophages, *Mol Microbiol*, **85**, 492-512.

5 Stern, A. M., Hay, A. J., Liu, Z., Desland, F. A., Zhang, J., Zhong, Z. & Zhu, J. (2012), The NorR regulon is critical for Vibrio cholerae resistance to nitric oxide and sustained colonization of the intestines, *MBio*, **3**, e00013-00012.

6 Stevanin, T. M., Moir, J. W. & Read, R. C. (2005), Nitric oxide detoxification systems enhance survival of Neisseria meningitidis in human macrophages and in

nasopharyngeal mucosa, *Infect Immun*, **73**, 3322-3329.

7 Karlinsey, J. E., Bang, I. S., Becker, L. A., Frawley, E. R., Porwollik, S., Robbins, H. F., Thomas, V. C., Urbano, R., McClelland, M. & Fang, F. C. (2012), The NsrR regulon in nitrosative stress resistance of Salmonella enterica serovar Typhimurium, *Mol Microbiol*, **85**, 1179-1193.

8 Robinson, J. L. & Brynildsen, M. P. (2013), A kinetic platform to determine the fate of nitric oxide in Escherichia coli, *PLoS computational biology*, **9**, e1003049.

9 Robinson, J. L., Miller, R. V. & Brynildsen, M. P. (2014), Model-Driven Identification of Dosing Regimens that Maximize the Antimicrobial Activity of Nitric Oxide, *Metabolic Engineering Communications*, **accepted**.

Re-Engineered β-Oxidation Reversal for the Synthesis of ω-Functionalized Products in Escherichia coli

Cheong, S., Rice University

Title:

Re-engineered Î²-oxidation reversal for the synthesis of Ï?-functionalized products in *Escherichia coli*

Abstract:

Due to its potential for high product yields, the newly developed Î²-oxidation reversal pathway is an efficient platform for the production of a wide range of industrially important fuels and chemicals. This pathway has been demonstrated to produce fatty acids and alcohols of varying chain length through the utilization of acetyl-CoA as the primer with termination by thioesterase or acyl-CoA reductase/alcohol dehydrogenase respectively. While these product classes contain many important fuels and chemicals, they are limited to compounds with a non-functionalized omega end as a result of using acetyl-CoA as the primer. Diversification of product classes with varying functionality can be achieved through the use of various functionalized priming molecules. The modular nature of a Î²-oxidation reversal can be exploited for this purpose through the use of core enzymatic pathway components along with activation and termination enzymes which are shown to function with Ï?-functionalized intermediates of interest.

The work reported here demonstrates the re-engineering of the reverse Î²-oxidation pathway to enable the utilization of Ï?-functionalized primers and synthesis of Ï?-functionalized products through identifying and utilizing enzymes capable of functioning with required intermediates. This was demonstrated through the production of adipic acid, a six-carbon Î±, Ï?-diacarboxylic acid with high industrial importance, using Ï?-carboxylated succinyl-CoA as the primer. The re-designed pathway consisting of selected core enzymatic modules for condensation between succinyl-CoA and acetyl-CoA and subsequent dehydrogenation, dehydration, and reduction steps resulted in adipic acid production *in vivo* when combined with suitable activation enzymes for the supply of succinyl-CoA and thioesterases for termination. Adipic acid titer of up to 0.25 g/L was achieved after further deletions of competing pathways. In addition to adipic acid, we are exploiting this re-engineered Î²-oxidation reversal for the synthesis of a wide-range Ï?-functionalized products.

Medium-Chain Carboxylic Acid and Alcohol Production Via Engineered β-Oxidation in E. coli

Kim, S., Rice University
Gonzalez, R., Rice University

Rising fossil fuel prices along with environmental concerns associated to their use have stimulated the development of alternative methods for fuel and chemical production from renewable resources. Previous studies have demonstrated the feasibility of producing aliphatic hydrocarbons derived from fatty acid biosynthesis in *Escherichia coli*. In this work, we present new methods to produce saturated and unsaturated carboxylic acids and alcohols using an engineered b-oxidation reversal. To increase the availability of acyl-CoAs for product synthesis, we deleted known endogenous thioesterases. Chromosomal over-expression of key core b-oxidation reversal and termination enzymes in this thiosterase-deficient strain led to saturated and unsaturated carboxylic acid production at titers reaching 0.5 g/L and 0.8 g/L, respectively. By combining the engineered b-oxidation reversal with appropriate acyl-CoA reductases, we achieved the direct production of medium-chain (C6-C10) alcohols from glycerol.

Cloning and Expression Profiling of Polycomb Gene, Vernalization Insensitive 3 from Tomato Solanum Lycopersicum L

Almutairi, Z.,

In many plant species, exposure to a prolonged period of cold during the winter promotes flowering in the spring, a process termed vernalization. VERNALIZATION INSENSITIVE 3 (VIN3) is a chromatin remodelling protein that is induced by low temperatures and is required for the vernalization response in *Arabidopsis thaliana.* VIN3 is one of the PcG proteins, which mediates epigenetic repression of FLOWERING LOCUS C (FLC) in *A. thaliana*. Here, we present cloning, characterization and expression of a putative *SlVIN3* gene in tomato (*Solanum lycopersicum* L.) by isolating cDNA clones corresponding to *SlVIN3* gene using primers designed based on conserved sequences between PcG genes in *A. thaliana* and tomato. The *SlVIN3* cDNAs were cloned into pBS plasmid and sequenced. Both 5' and 3' RACE were generated and sequenced. The flcDNA for *SlVIN3* gene of 2,823 bp length was composed of 5'UTR (336 bp), ORF (2,217 bp) and 3'UTR (270 bp). The translated ORF encodes a polypeptide of 739 amino acids. Alignment of deduced amino acids indicated that there are highly conserved regions between tomato *SlVIN3* predicted protein and plant VIN3 gene family members. Both unrooted phylogenetic trees constructed using maximum parsimony and maximum likelihood methods indicated that there is a close relationship between *SlVIN3* predicted protein and VIN3 protein of *Vitis vinifera.* The expression of *SlVIN3* remained high during floral organ differentiation and growth and decreased to its lowest level when the fruit starts to develop.

Engineering, Predicting, and Understanding Cofactor Specificity in Ketol-Acid Reductoisomerases

Cahn, J. K. B.,
Brinkmann-Chen, S.,
Arnold, F. H., California Institute of Technology

The branched-chain amino acid biosynthesis pathway has attracted considerable interest for the production of these valuable nutrients, and also as a source of potential biofuels. However, the NADPH-dependence of the enzyme which catalyzes the second step, ketol-acid reductoisomerase (KARI), has been a barrier to the large-scale fermentative implementation of this pathway. After successfully engineering NADH-preference in *E. coli* KARI by directed evolution, we developed a general recipe for engineering it in any KARI, and demonstrated it on a representative set. Subsequently, we used the insight gained from these engineering efforts to analyze thousands of experimentally uncharacterized KARI genes and predict ones that could utilize NADH. From these, we expressed and characterized four KARIs that preferred NADH and four that were bispecific for the two cofactors, the first such enzymes in a class previously believed to be exclusively NADPH-dependent. Crystal structures of two of these enzymes show insertions into the specificity-determining β2αB-loop, giving insight into the molecular evolution of cofactor specificity.

A Click Chemistry Approach to Site-Specific Immobilization of a Small Laccase Enables Efficient Direct Electron Transfer in a Biocathode

Chen, Z., Texas A&M University
Guan, D., Texas A&M University

Enzymatic biofuel cells and biosensors critically depend on efficient electron transfer between immobilized redox enzymes and the electrode. For direct electron transfer (DET) electrodes, the distance between the enzymatic active site and the electrode dictates electron harvesting efficiency, with empirical studies showing an optimum distance of approximately 20 Å. Since most proteins are much larger than 20 Å, strategic orientation of the enzyme on the electrode is crucial. In this study, we sought to enable site-specific electrode immobilization of a model small laccase from *Streptomyces coelicolor* (SLAC) through the *in vivo* incorporation of a non-natural amino acid with enabling chemistry, 4-azido-L-phenylalanine (AzF), into different sites in the vicinity of the catalytic center. AzF-containing laccase variants were immobilized onto a cyclooctyne-functionalized multi-walled carbon nanotube electrode via a copper-free cyclooctyne-azide cycloaddition reaction. One of the AzF-containing variants transferred electrons to the electrode with an efficiency, 28.7%, approaching the maximum possible within system-imposed constraints. The catalytic current generated was stable, with only ~14% reduction in current following 8 days at ambient temperature. This report provides a facile strategy for the site-specific directional immobilization of enzymes on electrodes for biofuel cell and biosensor development.

Rational Design of Antibody Cocktails to Treat Disease Caused By Bordetellae

Maynard, J. A., The University of Texas at Austin
Nguyen, A.,
Wagner, E. K., The University of Texas at Austin
Acquaye-Seedah, E.,
Wolf, R.,
Harvill, E.,
Papin, J.,

In spite of near-universal vaccination, pertussis (aka whooping cough) remains a significant public health problem with no specific therapy for established disease. Because disease is not driven by bacterial infection at the time pertussis is diagnosed, antibiotic therapy is often ineffective. High-titer polyclonal antibody preparations (IVIG) have been previously evaluated in mice and humans, with demonstrated ability to protect mice up to seven days after infection but less clear outcomes in infants. We hypothesize that cocktails of anti-toxin antibodies, selected to bind key protective epitopes with high affinity will be more effective at ameliorating disease in a passive immunotherapy setting and these results can guide design of next generation vaccines.

To develop a potential passive immunization therapy, we turned to two well-characterized pertussis-toxin binding monoclonal antibodies developed by the Sato group in the 1980s, 1B7 and 11E6. We have characterized their mechanisms of action, which explains their observed synergy: 11E6 inhibits the toxin-receptor interaction, while 1B7 can bind receptor-bound toxin and re-direct intracellular trafficking of the toxin. These murine antibodies have been humanized via exclusively *in silico* methods and expressed with human immunoglobulin constant regions in CHO cells Full biochemical characterization shows high binding affinity, increased stability and high *in vitro* toxin neutralizing activity. The humanized antibodies were evaluated in a mouse model in which the animals were infected with a virulent pertussis strain isolated from a critically ill newborn. The antibodies, individually and in combination, enabled the mice to continue to gain weight, diminished bacterial colonization in the lungs, and completely mitigated the rise in white blood cell count (WBC) that is characteristic of disease. Importantly, they were superior to the IVIG previously used in human clinical trials. We have further evaluated the antibodies' ability to protect baboons when administered on day 3 after infection when the white blood cell count is elevated. Here, antibody-treated but not control animals saw a rapid decrease in WBC, coughing and a modest decrease in bacterial colonization.

We conclude that these antibodies are protectiv ein both animal models of disease and that they may be protective in humans as well. We have initiated GMP manufacturing processes in anticipation of IND filing in 2015.

Production of Fatty Acid-Derived Fuels and Chemicals in Saccharomyces Cerevisiae

dEspaux, L.,
Keasling, J. D., Lawrence Berkeley National Laboratory
Runguphan, W.,

Rapidly increasing energy demands, limited fossil fuel reserves, and the environmental risk associated with extracting these have led to a growing interest in microbial biofuel production. Fatty acids, used in nature for cellular energy storage, are among the most energy-dense molecules produced naturally, and can be converted into high-value molecules such as fuels, lubricants, and other chemicals. We previously reported on engineered the budding yeast *Saccharomyces cerevisiae* to produce free fatty acids at titers up to 400 mg/L. Here, we systematically optimized our original production strain by optimizing flux through the fatty acid biosynthesis pathway, balancing cofactor recycling, limiting production of side products, and optimizing the metabolic state of the cell. We combined beneficial engineering strategies to produce a fatty acid overproducer with production titers over 800 mg/L. Our results demonstrate the power of rational, systematic engineering to overcome pathway inefficiencies and improve product titer.

Improving Cell-Free Protein Synthesis through Genome Engineering of Escherichia coli Lacking Release Factor 1

Martin, R. W., Northwestern University
Hong, S. H., Northwestern University
Kwon, Y. C., Northwestern University
De Soye, B.,
Ntai, I., Northwestern
de Paz, A.,
Kelleher, N.,
Jewett, M. C., Northwestern University

Site-specific incorporation of non-standard amino acids into proteins by amber suppression grants access to novel therapeutics and new types of biomaterials. Here, we describe the development of a high yielding cell-free protein synthesis (CFPS) platform from crude extracts of genomically recoded *Escherichia coli* lacking release factor 1 (rEc.E13.D*prfA*). We exploited genome engineering to design and construct synthetic genomes that upon cell lysis lead to improved extract performance. We targeted the deletion of five potential negative effectors, the nuclease genes *rna*, *rnb*, *csdA*, *mazF*, and *endA*. Our most productive extract, resulting from a *csdA* and *endA* double mutant, synthesized 550 ± 40 μg/mL of modified superfolder green fluorescent protein containing p-acetyl-L-phenylalanine and achieved ~84% suppression efficiency. The modified protein yield was further improved to 1300 ± 100 μg/mL using a semi-continuous cell-free reaction format, which is the highest reported for extracts derived from RF1-deficient strains. Our work has implications for using whole genome editing for CFPS strain development and cell-free synthetic biology.

Constructing a Synthetic Metabolic Pathway Protects Escherichia coli Against the Acetic Acid Inhibition and Fulfills Enantiomerically Pure (R, R)-2,3-Butanediol Production

Ji, X., Nanjing Tech University
Huang, H., Nanjing Tech University

Enantiomerically pure (R, R)-2,3-butanediol has unique applications due to its special chiral group and spatial configuration. Currently, its chemical production route has many limitations. In addition, no native microorganisms can accumulate (R, R)-2,3-butanediol with an enantio-purity over 99%. Herein, we constructed a synthetic metabolic pathway for enantiomerically pure (R, R)-2,3-butanediol biosynthesis in *Escherichia coli*. The fermentation results suggested that introduction of the sythetic metabolic pathway redistributed the carbon fluxes to the neutral (R, R)-2,3-butanediol, and thus protected the strain against the acetic acid inhibition. Additionally, it showed that the traditionally used isopropyl beta-D-thiogalactoside (IPTG) induction displayed negative effect on (R, R)-2,3-butanediol biosynthesis in the recombinant *E. coli*, which was probably due to the protein burden. With no IPTG addition, the (R, R)-2,3-butanediol concentration reached 115 g/L by fed-batch culturing of the recombinant *E. coli*, with an enantio-purity over 99%, which is a new record of enantiomerically pure (R, R)-2,3-butanediol production.

A Reusable Immobilized Intein Segment That Eliminates Premature Cleaving and Thiol Requirements in Intein-Based Protein Purification

Shi, C., Ohio State University
Cain, D., The Ohio State University
Wood, D. W., The Ohio State University

Inteins are self-contained splicing protein elements, which can be modified to generate self-cleaving tags for recombinant protein purification. Although intein-based purification tags have been in use for over 15 years, they are still hampered by premature cleaving *in vivo* during protein expression. A recent advance has been the introduction of the naturally split *Npu* DnaE intein. This intein can be expressed in two separate segments, each of which is inactive for splicing or cleaving, but exhibits very rapid assembly and splicing when the segments are combined. A cleaving version of this ultra-fast intein promotes full cleavage and release of a purified target protein in less than half an hour at room temperature. In an approach developed by Zhilei Chen and coworkers, the N-terminal *Npu* intein segment is fused to an affinity tag and immobilized onto a conventional affinity resin, while the C-terminal intein segment is fused to the target protein for purification. Because the C-terminal intein segment is inactive for cleaving, this system allows high overexpression of the tagged target protein without significant premature cleaving. On-column assembly of the intein segments is then used to purify the fused target protein, followed by addition of 50 mM dithiothreitol (DTT) to induce rapid C-terminal cleavage and release of the target protein. Unfortunately, the necessity for a high concentration of DTT makes this approach problematic for proteins containing disulfide bonds, while the use of a conventional affinity tag can lead to leaching of the N-terminal segment and assembled intein from the column during purification. In the present work, we have solved these problems by reengineering the existing *Npu* DnaE split intein system. Specifically, we have eliminated all cysteine residues within the intein through site-directed mutagenesis, and we have covalently immobilized the intein N-terminal segment to Sulfolink® agarose resin. The covalent immobilization of the intein N-terminal segment provides a reusable capture resin for split-intein purifications, which is more stable than a conventional affinity tag coupling and can be regenerated for multiple cycles of use. Further, the covalent linkage and strong intein segment association provide higher purity of the recovered target; typically over 90% based the PAGE image analysis. More importantly, reducing agents are no longer necessary to induce a rapid cleaving reaction, making this method safe for proteins that contain disulfide bonds. We have fully characterized the cleaving kinetics of the engineered intein system under different conditions, including temperature, initial target protein amino acid and pH. We have also demonstrated this system with several targets expressed in bacterial and mammalian cell-free expression systems. We anticipate that the reengineered split intein system, with the resulting reusable capture resin, has the potential to provide a highly useful purification platform for recombinant proteins expressed in virtually any host cell system.

Design of Molecular Biosensors for New or Improved Polyketide Synthase Activities

Cirino, P. C., University of Houston
Wang, Z.,

Metabolite-responsive gene regulators are valuable tools in synthetic biology and can be used as molecular reporters for protein and metabolic engineering. To develop biosensors for high-throughput screening of biosynthesis pathways in *E. coli*, we sought variants of the AraC regulatory protein that respond to small molecules of interest. One of our most effective reporters from these efforts responds to the compound 4-hydroxy-6-methyl-2-pyrone, or triacetic acid lactone (TAL), commonly identified as a triketide derailment product during polyketide biosynthesis. Our AraC-based TAL reporter (AraC-TAL) enabled the in vivo directed evolution of *Gerbera hybrida* 2-pyrone synthase activity in *E. coli* [Tang et al., J. Am. Chem. Soc. 2013, 135, 10099] and is now being used to evolve variants of a fungal polyketide synthase (PKS) that functionally express in *E. coli*.

Natural polyketides and their ?unnatural? analogues offer great potential as therapeutics, including antibiotics, antifungals, and anticancer agents. Extensive research is aimed at modifying PKS substrate and product profiles, as means to better understand PKS biochemistry and to broaden chemical diversity. However, directed evolution of these enzymes is limited by a general lack of product-specific, high-throughput screening assays. By exploiting the promiscuity of select PKS domains, a variety of triketide lactones possessing the 4-hydroxy-2-pyrone moiety have been synthesized [Winter et al., Organic Letters. 2013, 15, 3775; Koryakina et al., Org. Biomol. Chem. 2013, 11, 4449]. We probed the structural specificity of AraC-TAL and found that this reporter responds to a variety of 4-hydroxy-2-pyrones to different extents, and not to similar compounds but lacking these features. Structural and mutational analyses were then carried out to better understand how AraC-TAL and variants thereof distinguish between select 2-pyrones. This presentation highlights results from these studies, as well as progress toward the design of AraC-TAL variants specific for select TAL derivatives, which will facilitate PKS directed evolution towards synthesis of the respective triketides and their corresponding ?unnatural? products resulting from complete PKS processing.

Modular Pathway Optimization for Production n-Butanol from Red Seaweed in Escherichia coli

Lim, H. G., POSTECH
Lim, J. H., POSTECH
Jung, G. Y., POSTECH

Marine biomass, known as the third generation of biomass, has many advantages rather than conventional biomass due to its high biomass productivity, carbon fixation rate, and easy pretreatment process, etc. Especially, red seaweed which is mainly cultivated in Asia, contains galactose and glucose, many bio-chemicals can be produced from these sugars through fermentation process. However, the challenging problem is that galactose is not favored by many industrial microorganisms (for example, *Escherichia coli*). Slow utilization rate and carbon catabolite repression limits carbon flux and results in low productivity and titer. In this research, we introduced two synthetic modules to utilize sugars from hydrolysate of red seaweed (*Gelidium amansii*) and to produce n-butanol, one of advanced fuel. The modules with metabolic enzymes under synthetic promoters, rationally designed 5'UTRs (5'untranslated region), and terminators for maximum catalytic activity were assembled, and these modules were optimized by finding maximum productivity with various expression of specific gene. With this approach, the engineered strain showed significantly enhanced sugar utilization and n-butanol production from both red seaweed mimetic media and real hydrolysate. Finally, we could confirm the possibility of marine biomass as carbon source to produce many valuable bio-products instead of fossil fuel.

Metabolic Engineering of Escherichia coli for Production of L-Tryptophan Using Synthetic Biology Tools

Jang, S., Pohang University of Science and Technology
Seo, S., Pohang University of Science and Technology
Jung, G. Y., POSTECH

Current metabolic engineering is limited by the ability to finely and predictively express pathway enzymes and to screen superior and optimized variants from library. Here, we present synthetic biology tools which settled above problems to produce L-tryptophan from *Escherichia coli*. First, we amplified metabolic flux of L-tryptophan pathway using a predictive model for gene expression level, UTR Designer, that can precisely predict expression level of gene based on folding energy of specific features in mRNA secondary structure. UTR sequences of bottleneck enzymes of the pathway were re-designed. Then we optimized the producing strain using a riboswitch-based screening tool, riboselector, that can modulate expression level of selective marker gene in response to the target molecule. Riboselector for L-tryptophan was able to enrich superior variants from vast library which were constructed from rationally engineered strain. We claim here that high-performance producing strains could be engineered using novel synthetic biology tools.

Modular Ligand-Activated Enzymes from Antibody Mimetic Proteins

Nicholes, N., Johns Hopkins University
Date, A., Johns Hopkins University
Beaujean, P., Johns Hopkins University
Kanwar, M., Johns Hopkins University
Hauk, P.,
Ostermeier, M., Johns Hopkins University

Ligand-activated enzymes (protein switches) can be constructed by the correct fusion of ligand-binding domains and enzymes. For example, maltose-activated beta-lactamases have been constructed by the insertion of beta-lactamase (BLA) into the maltose binding protein (MBP) from *E. coli*. In these protein switches, BLA catalytic activity is a function of maltose concentration. The process of developing a new switch is time consuming and labor intensive. To accelerate our ability to construct switches that are activated by new protein and small molecule targets, we have worked to develop modular platforms for protein switches using monobody and DARPin antibody mimetics as input domains. The advantage of these proteins is that they have been engineered to bind with high affinity to wide variety of protein targets. In contrast to the periplasmic binding proteins, however, these antibody mimetic proteins do not undergo a large conformational change on ligand binding. First, MBP-activated beta-lactamases were created by a directed evolution approach involving the fusion of BLA with the MBP-binding DARPin off7 and the fusion of BLA with the MBP-binding monobody MBP-74. We identified switches that function both *in vitro* and in *E. coli* cells. Next, these MBP-activated switches were converted for activation by new protein targets from bacteria, yeast and other eukaryotes by transplanting mutations into the antibody mimetic domain known to cause binding to new targets. The modularity of these protein switches will be discussed along with the implications for their future use as potential biosensors.

Developing an Enzymatic Switch from HSV-Thymidine Kinase As a Potential Cancer Therapeutic

Shelat, N., Johns Hopkins University
Ostermeier, M., Johns Hopkins University

Developing an enzymatic switch from HSV--thymidine kinase as a potential cancer therapeutic.

Nirav Y. Shelat and Marc Ostermeier

The development of ligand--activated protein switches holds tremendous potential for protein--based therapeutics.A specific application of engineered switches is in gene directed enzyme--prodrug therapy (GDEPT).GDEPT discriminately kills cancer cells through targeted DNAdelivery of a suicide gene, but is limited by the specificity of gene delivery methods.An engineered protein switch would be able to add another level of specificity towards cancer cells beyond that of the DNA delivery method and make GDEPT a more attractive therapeutic strategy. To this end, we have developed a directed evolution approach for creating a protein switch from the herpes simplex virus thymidine kinase (HSV--TK), an enzyme often used in GDEPT with the prodrug ganciclovir. The desired switch would couple the intrinsic HIF1a-- binding affinity of the CH1 domain from the p300 protein to the activity of the HSV-- TK enzyme, creating a HSV--TK protein switch that is active only when HIF1--a is available. Since HIF1--a is a cancer marker, such a switch would cause the activation of the prodrug, ganciclovir, into ganciclovir monophosphate, specifically in cancer cells. Our strategy to make such switches involved the generation of a DNA library that had variants with a CH1 domain inserted in various locations in the HSV--TK gene. A two tier selection scheme was implemented to then select for fusions which possessed the desired ?OFF? and ?ON? states. Upon screening, we identified CH1-- HSV--TK gene fusions that conferred a switch phenotype to E.coli cells that is dependent upon the induced expression of HIF1--a. These constructs will serve as a basis for future rounds of directed evolution that will further increase the activity window between the ?ON? and ?OFF? states of the HSV--TK protein switch

Heterologous Reconstruction of a Synthetic Bacterial Oscillator
Chen, A. H.,

ABSTRACT Submission for Oral Presentation

Heterologous Reconstruction of a Synthetic Bacterial Oscillator

Anna H Chen.[1,2], David Lubkowicz,[1,2], Vivian Yeong[1,2], Roger L. Chang[1,2], Pamela A. Silver[1,2].

[1]Department of Systems Biology, Harvard Medical School, Boston, MA USA;

[2]Wyss Institute for Biologically Inspired Engineering, Boston, MA USA

Many synthetic biological circuits have been constructed to enhance production of high value compounds via spatial organization or logic gates. However, only a limited number of temporal control systems are currently available and biologically-relevant synthetic timing modules are non-existent.

The circadian clock in cyanobacteria controls global gene expression. The phosphorylation state of KaiC oscillates, facilitated by interactions with KaiA and KaiB. These changes are thought to mediate circadian rhythms in global transcription via output factors SasA, RpaA, RpaB, and CikA; however, the exact role of these output factors is an area of active research.

In order to construct a modular synthetic circadian system and to study the native clock independent of other intrinsic factors, we heterologously expressed the cyanobacterial circadian clock genes and output factors in *E. coli*. We show that heterologous expression of the cyanobacterial clock proteins in

E. coli results in phosphorylation state changes of KaiC over a 24-hour period. This enabled us to systematically characterize the function of the output factors that transmit circadian information from KaiC to a transcriptional output. Furthermore, using an interaction-dependent transcriptional system with a fluorescent reporter, we show circadian oscillations in *E. coli*. Taken together, these results indicate that the bacterial circadian oscillator is modular and can be engineered in a heterologous organism *in vivo*. An engineered circadian clock not only allows us to better understand the physiology of the clock, but also can be incorporated in complex synthetic circuits that may have chronotherapeutic and industrial applications.

Synthetic Biology Approach to Systematic Optimization of Protein Function

Govindarajan, S.,

Current gene synthesis technologies allow precise control of all sequence features critical for biological function. Variables at the levels of amino acid substitutions, vector elements and host genome mutations can all be critical for achieving the desired output. Modern machine learning tools and algorithms are well equipped to identify and quantify the relative contributions of each variable. We can address questions related to additivity and multi-dimensional effects of substitutions on various properties and activities that can be measured accurately under commercially viable condition.

We present data and models from several projects illustrating the advantages of this approach.

References:

ACS Synth Biol 2014. Mapping of Amino Acid Substitutions Conferring Herbicide Resistance in Wheat Glutathione Transferase. Govindarajan et al.

J Am Chem Soc 2013. Improved biocatalysts from a synthetic circular permutation library of the flavin-dependent oxidoreductase Old Yellow Enzyme. Daugherty et al.

Protein Eng Des Sel 2013 26(1):25-33. Redesigning and characterizing the substrate specificity and activity of Vibrio fluvialis aminotransferase for the synthesis of imagabalin. Midelfort, KS. et al.

Design and Characterization of Synthetic, Minimal Promoters in Yeast

Redden, H., Molecular Biosciences
Alper, H., The University of Texas at Austin

Promoters are critical to successful engineering in the eukaryotic production platform *Saccharomyces cerevisiae*. For decades, native promoters and their mutants have been used for metabolic engineering and synthetic biology efforts. However, native promoters are limited in expression range and vulnerable to cellular regulation, and mutants of these promoters inherently have high sequence homology. Moreover, these promoters are bulky elements--usually hundreds of base pairs long. To address these issues, we sought to develop non-homologous, purely synthetic, minimal promoters for yeast. To do so, we apply a hybrid promoter approach to a unique collection of library-derived promoter elements. This set includes constitutive and inducible promoters driving expression as strong as native promoters, but in less than 1/3 of the DNA sequence. In doing so, we generate the shortest promoters described to date and develop a set that shows minimal context specificity. This work builds a framework for designing modular, minimal promoters with desired expression traits.

Introduction of Heterologous Mevalonate Pathway in Methylobacterium Extorquens AM1 for High Production of Value-Added Compounds from Methanol

Zhu, W. L., Tsinghua University
Zhang, C., Tsinghua University
Cui, L. Y., Tsinghua University
Xing, X. H., Tsinghua University

Industrial biotechnology plays an important role in meeting growing demand for green economy. However the feedstock being used is mainly sugar, which limits the development of industrial biotechnology in terms of relative high price, seasonal fluctuation of supply of feedstock and policy restriction on guarantee for food. Methylotrophic bacteria, which utilize reduced carbon substrates (such as methane, methanol, and other methylated compounds) as sole sources of carbon and energy, provide a potential way to solve such problems under the glut of methane and methanol. To produce desirable value-added products by methylotrophic synthesis, we focused on mevalonate pathway, from which about 50,000 terpenoid compounds in nature are derived, including artemisinin, paclitaxel, isoprene, etc.

By introducing heterologous 3-Hydroxy-3-methylglutaryl-coenzyme A (HMG-CoA) synthase (HMGS) and HMG-CoA reductase (HMGR) into *Methylobacterium extorquens* AM1, the top portion of mevalonate pathway can act in this strain, and convert acetyl-CoA to mevalonate acid. AM1-mve consisting of both HMGS and HMGR from *Enterococcus Faecalis*; AM1-mvh consisting of HMGS from *Blattella Germanica* and HMGR from *Trypanosoma cruzi* (either of which showed the highest enzymatic activity compared to other species), were first constructed. In flask cultivation, mevalonate production of AM1-mve was 0.30g/L, and that of AM1-mvh was 0.10g/L. Heterologous acetoacetyl-CoA thiolase was then introduced into AM1-mve and AM1-mvh respectively to construct engineered strains AM1-mva and AM1-mvt. With overexpression of acetoacetyl-CoA thiolase, mevalonate production of AM1-mvt increased 2.5 times over AM1-mvh, from 0.10g/L to 0.35g/L. However, AM1-mva showed a little decrease on production than AM1-mve, from 0.30g/L to 0.25g/L, and appeared severe cell clumps. Such results showed that acetoacetyl-CoA thiolase played a key role in balancing and broadening metabolic flux of heterologous mevalonate pathway. Therefore, 6 engineered strains carrying a series of plasmids with different RBS strengths in front of heterologous *phaA*gene (encoding acetoacetyl-CoA thiolase) of AM1-mvt were further constructed. With different levels of acetoacetyl-CoA thiolase expression, those strains showed significant differences on mevalonate production. By fermenting the best performed strain, a production of 3.2g/L in a 3L-bioreactor batch was achieved.

This study was the first time that mevalonate pathway was successfully expressed in methylotrophic strains. With the engineered chassis strain, we can easily synthesize desirable value-added terpenoids by introducing some heterologous gene further, which bright the future of large-scale production of value-added compounds from methanol.

Acknowledgement
National Natural Science Foundation 21376137

Antibody Mediated Protection in Whooping Cough Infection Due to Bordetella Pertussis

Acquaye, E.,
Maynard, J., University of Texas at Austin

Incidence rates of whooping cough, a highly infectious disease caused primarily in humans by the bacteria, *Bordetella pertussis (B. pertussis)*, decreased dramatically as result of whole cell pertussis (wP) immunizations. Concerns about the safety of the wP vaccines resulted in the development of new vaccines using acellular components of *B. pertussis*. In spite of widespread vaccine availability, whooping cough continues to affect infants and young children, often with serious consequences. Disease rates in developing countries have risen steadily over the last few decades to 42,000 cases in 2012. The safer acellular pertussis (aP) vaccines prevent the severe manifestations of the disease but are unable to completely eliminate subclinical infection.

In order to design improved vaccines, it is important to understand the mechanisms which induce protective immunity to the pertussis bacteria. My research work aims to better understand the biochemical attributes of antibodies produced after humans receive an aP vaccination. Of the numerous *B. pertussis* antigens, the pertussis toxin (PTx) and filamentous hemagglutinin (FHA) have been identified as major factors which elicit a humoral response sufficient for protection. After booster immunization with an aP vaccine, representatives of monoclonal antibodies responding to PTx or FHA were expressed and purified. The affinity and neutralizing ability to the antigens were also measured and the epitopes recognized are currently being classified. The results of this study are expected to expand the present knowledge of the molecular mechanisms of antibody mediated protection after aP vaccination and improve the design of the vaccines. The study also identifies antibody candidates which neutralize PTx or FHA and maybe useful for passive immunotherapy of whooping cough.

New Biophysical Model for Characterization of Human Heart Valves

Garcia, J., Universidad de Monterrey
Padilla, J.,
Rivera-Silva, G., Universidad de Monterrey
Moreno, M., Universidad de Monterrey

New Biophysical Model for Characterization of Human Heart Valves

Juan B. Garcia Flores[1], Jose C. Padilla Jaime[1], Gerardo Rivera Silva[1] & Maria G. Moreno Trevino[2]

[1]Laboratory for Tissue Engineering and Regenerative Medicine, University of Monterrey, San Pedro Garza Garcia, NL, MEX
[2]Basic Sciences Department, University of Monterrey, San Pedro Garza Garcia, NL, MEX

Background. Utilizing human valves often means less risk of infection, even though it is a more complicated surgery than other replacement techniques. Homograft rescue should be realized within 48 hours post mortem. Appropriate homografts are preserved either under 0--4oC for further use within 6 weeks after rescue, or are frozen at --150oC. However, an integral morphological examination is not performed before the surgery. Methods. We developed a morphological approach to show the cell geometric interaction and valvular integrity. Microscopic techniques and photographs were used to measure angles and lengths (beta 4.03 of Scion Image Software). Three groups were established: I (aortic valves; n= 40); II (pulmonary; n= 38); and III (mitral; n= 38). The shape prediction was measured using: M(T)´= T--1TPS (X(T)). Depending on the situations, we used one--way repeated--measures ANOVAS. Results. The angles related with valvular integrity and cardiac CT means were: 47± 1o , 1.34 ± 0.40 mm (aortic); 35±2o, 1.99 ± 0.36mm (mitral); and 60 ± 2o, 1.57 ± 0.30 (pulmonary). These results reported an interaction between geometric expression and valvular integrity in all groups. These approaches can theoretically improve the management of individuals affected by valvular heart disease, diminishing risks for problems during surgical procedures.

Reengineering the Genetic Code to Expand the Language of Biology Using a Cell-Free Synthetic Biology Approach

Bundy, B. C., Brigham Young University
Smith, M. T., Brigham Young University
Hunt, J., Brigham Young University
Salehi, A., Brigham Young University
Broadbent, A., Brigham Young University
Schinn, M., Brigham Young University

The incorporation of non-standard amino acids (nsAAs) is a protein engineering game changer that is transforming our ability to make better therapeutics, medical diagnostics, and biocatalysts. However, as we begin to explore the potential of incorporating more than one nsAA simultaneously into our protein of interest, we are limited by the genetic code. While theoretically only 21 codons are necessary to incorporate the natural proteogenic cannon of 20 amino acids, all of the 64 natural codon possibilities have a corresponding tRNA or release factor. Hence there is a need to reengineer the genetic code to emancipate codons for nsAA use beyond the popular amber stop codon target. Here report the application of a cell-free synthetic biology approach toward this objective.

A cell-free approach is well suited toward our application due to the direct access afforded by the system. This access enables rapid optimization and unprecedented control of both the endogenous protein synthesis machinery and the exogenous machinery necessary for nsAA incorporation. In addition, the cell-free system is not limited by the transport or toxicity of non-standard products and intermediates (e.g. nsAA, exogenous machinery necessary for nsAA incorporation). Finally, as we consider transfer to commercial applications, the cell-free system has been demonstrated to be reliably and easily scaled from the laboratory scale to the commercial scale.

A Novel Storage Method for Umbilical Cord Blood

Afshan, G., Milwaukee School of Engineering
Hassemer, E., Milwaukee School of Engineering

NoveStororagMeethod Â fUmmbilicCord Â d Â Blood

Afshan, Gul S; Hassemer, Eryn L

Currently, umbilical cord blood (UCB) is cryopreserved for medical applications by private and public banks with high initial processing and annual storage fees. This can be a significant financial burden, therefore, a new method is currently under way to store UCB. Historically, lyophilization of mammalian cells yields low cell viability upon rehydration. We are pursuing an alternative lyophilization protocol that uses a more natural cryoprotectant derived from a plant sugar and green tea. This protocol uses a newly designed millifluidic device that serves as both a sample chamber and long-term storage vessel for UCB. It is capable of withstanding the extreme conditions that occur during lyophilization and autoclaving. As the cell viability is directly related to the recovery process, it is projected that higher viability can be achieved for cells stored and rehydrated in the millifluidic device. This technology is being developed to enhance the availability of UCB banking for families from all economic ranks by eliminating the annual fees for liquid nitrogen, maintenance, and storage space. The compact design of the millifluidic device would enable storage under less stringent conditions. This project offers an innovative, marketable solution for utilizing umbilical cord blood stem cells for a growing number of therapeutic uses. The work has been initiated as a senior design project and continues with further development of the prototype along with the methods and safety processes associated with it.

Design of Halotolerant Bovine Dnase Guided By Phylogenetic and Domain Structure Studies of Bacterial Dnases

Alzbutas, G., Vilnius University
Kaniusaite, M., Thermo Fisher Scientific Baltics
Grybauskas, A.,
Lagunavicius, A.,

Analysis of sequence databases suggested that non - specific DNA binding domain is characteristic to many halophilic or halotolerant bacterial DNase species and it could serve as a mechanism to maintain enzyme activity at high ionic strengths.

This hypothesis was tested on one secreted bacterial DNase from halophilic bacterium Thioalkalivibrio sp., which was experimentally shown to maintain activity at almost saturated solution of sodium chloride. The enzyme is composed from two domains: N-terminal domain which exhibits DNase activity and C-terminal domain comprising duplicate HhH motifs, which natural functions are recognition and interaction with DNA.

We have experimentally proved, that the C-terminal domain in Thioalkalivibrio sp. DNase is responsible for enzyme halophilicity. Based on these findings we designed two variants of bovine DNase I C-terminal fusions: one with fused C-terminal domain of DNase from halophilic bacterium *Thioalkalivibrio sp.* , the second one with fused homologous domain from ComEA protein of *B. subtilis*. These fusions were showed to be more salt tolerant than wild type DNase and subsequently superior to it at least in several practical applications

Composability and Design of Parts for Large-Scale Pathway Engineering in Yeast

Young, E. M., Massachusetts Institute of Technology
Roubos, J. A., DSM Biotechnology Center
Meijrink, B., DSM Biotechnology Center
Voigt, C. A., University of California, San Francisco

Yeasts can be powerful platforms for production of many industrially relevant molecules. However, expression systems for predictable control of gene expression do not permit the construction of large multi-gene combinatorial libraries without multiple repetitions of promoters and terminators, compromising construct stability.

Therefore, to enable design and construction of these libraries in yeast, it is necessary to develop assembly methods and parts capable of achieving the size and diversity needed to gather useful information concerning pathway design.

Yeast promoters are among the most heavily studied genetic elements, yet only a handful are in use industrially. Even fewer terminators are well understood and in use. Recent literature has demonstrated that terminators can have a profound effect on gene expression. Yet, it remains unknown if this effect is independent of the promoter. To this end, the variation in expression of GFP was measured for 244 promoter-terminator combinations. This value is compared to the strength measured using a standard promoter or terminator. These data can be applied to optimize expression of metabolic pathways, and the approach extended to characterize the behavior of new parts, natural or synthetic.

Cofactor Engineering for the Production of Short Chain Fatty Acid

HAN, S., Rice University
San, K. Y., Rice University

Cofactor engineering for the production of short chain fatty acid

SongI Han, Ka-Yiu San

Department of Bioengineering, Rice University, Houston, TX

Free fatty acids can be used as precursors for the production of chemicals and biofuels. *E. coli* possesses the biosynthetic fatty acid pathway and the introduction of acyl-ACP thioesterase enables *E. coli* to produce free fatty acid with specific chain length in a substantial amount. In particular, many challenges have been proposed when producing short chain length free fatty acid in *E. coli*, which might be attributed to the toxicity, severe feedback inhibition, metabolic flux imbalance and so on. Thus, the respective studies with regard to the strategies proven to be helpful in the case of long chain free fatty acid accumulation need to be implemented independently to figure out the principles raveled in short chain free fatty acid production. Among a variety of approaches, the control of ratios of reducing equivalents appears to be important in that there are two steps to be involved in requiring cofactors such as NADPH or NADH in the elongation cycle of fatty acid biosynthesis. Enzymatic approaches can be plausible in *E. coli* bioprocessing system as a way of changing the cofactor ratios. For example, UdhA or PntAB, soluble pyridine nucleotide transhydrogenase or energy-dependent membrane-bound transhydrogenase, respectively can contribute to improving NADPH availability. In addition, NADK is linked to regulating the concentration of intracellular NAD_+ and $NADP_+$. In this study, cofactor engineering with UdhA, PntAB, and NADK in terms of producing short chain length free fatty acid was explored in engineered *E. coli* with different copy number plasmids. Experimental results demonstrated the effectiveness of cofactor engineering in enhancing the accumulation of short chain length free fatty acid.

Photocrosslinking with p-Azidophenylalanine Used to Identify the Location of Cic Binding on ERK

Futran, A.,
Shvartsman, S. Y., Princeton University
Link, A. J., Princeton University

Extracellular signal-regulated kinase (ERK) is responsible for integrating cellular responses to a wide range of stimuli. ERK-mediated phosphorylation of its many substrates regulates critical cellular processes including proliferation, differentiation and death. The transcriptional repressor Capicua (Cic) is one such ERK substrate that was first identified in *Drosophila* and has been implicated in a number of human diseases including cancer and neurodegenerative disorders. The details of Cic recognition by ERK remain largely unknown. We employed a novel strategy using biophysical and chemical biological assays to dissect the interaction between ERK and Cic. Using pull-down assays and Biolayer Interferometry, we identified residues in Cic that are required for its interaction with ERK. We then incorporated the unnatural amino acid p-azidophenylalanine into Cic positions adjacent to these critical ERK-interacting residues. This enabled the covalent crosslinking of ERK and Cic at the binding interface by UV irradiation. Analysis of the proteolytically digested ERK/Cic crosslink product by LC-MS/MS enabled the identification of the precise location on ERK used to bind to Cic, providing a detailed picture of the protein-protein interaction between Cic and ERK.

In-Depth Determination and Analysis of the Human Paired Heavy and Light Chain Antibody Repertoire

DeKosky, B. J., University of Texas at Austin
Kojima, T., University of Texas at Austin
Rodin, A.,
Hoi, K. H.,
Lee, J., University of Texas at Austin
Charab, W.,
Ippolito, G. C., University of Texas at Austin
Ellington, A. D., University of Texas at Austin
Georgiou, G., The University of Texas at Austin

High-throughput immune repertoire sequencing has emerged as a critical step in the understanding of adaptive responses following infection, vaccination or in autoimmunity. However determination of native antibody variable heavy:light pairs (VH:VL) remains a major challenge and no technologies exist to adequately interrogate the $>10^6$ B cells in typical specimens. We developed a low-cost single-cell emulsion-based technology for sequencing antibody VH:VL repertoires from $>2x10^6$ B cells per experiment with demonstrated pairing precision >97%. A simple flow-focusing apparatus was used to sequester single B cells into emulsion droplets containing lysis buffer and magnetic beads for mRNA capture; subsequent emulsion RT-PCR generated VH:VL amplicons for NextGen sequencing. Massive VH:VL repertoire analyses of human donors enabled rapid antibody discovery of HIV broadly neutralizing antibodies and cross-reactive influenza antibodies while also providing novel immunological insights including the identity, frequency, and pairing propensity of shared, or "public" VL genes, and detection of allelic inclusion (an implicated autoimmune mechanism) in healthy individuals. As DNA sequencing technologies continue to progress, low-cost high-throughput single-cell antibody sequencing such as reported here may enable rapid antibody discovery and provide new insights into humoral immune responses associated with vaccine development, autoimmunity, infectious diseases, and other human disease states.

Integrative Characterization of Human Cellular and Serological Antibody Repertoires Elicited by Seasonal Influenza Vaccine

Lee, J., University of Texas at Austin
Boutz, D. R., University of Texas at Austin
Vollmers, C., University of Texas at Austin
DeKosky, B. J., University of Texas at Austin
Horton, A. P., University of Texas at Austin
Murrin, E. M.,
Park, D.,
Ippolito, G. C., University of Texas at Austin
Marcotte, E. M., University of Texas at Austin
Quake, S. R.,
Georgiou, G., The University of Texas at Austin

Vaccines convey protection by stimulating B cells to produce a diverse repertoire of antigen-specific antibody proteins for an extended period of time. While next-generation sequencing has become an invaluable tool for investigating the cellular response to influenza vaccination, a comprehensive delineation also requires an understanding of the serological repertoire through the proteomic determination of the monoclonal antibodies (mAbs) comprising the vaccine serum response. By integrating transcript- and protein-level repertoire analyses using methodologies recently developed by our laboratory (Lavinder *et al.*, PNAS 2014; DeKosky *et al.*, NBT 2013), we have characterized the mAbs that constitute polyclonal responses following the trivalent-inactivated 2011-2012 seasonal influenza vaccination (FluZone) in four healthy donors. Specifically, the cellular and serological IgG repertoire to each HA from three virus strains comprising the vaccine (H1, H3 and B) was determined at days 0, 28 and 180 following vaccination. Select key findings include: (1) Clonotypic diversity of B repertoires was significantly larger than H1 and H3 repertoires. (2) A significant portion of the serological response to H1 (ranging from 40% to 63%) also displays binding to H3. Recombinant expression and analysis of cross-reactive antibodies identified by the proteomic analysis of the serum revealed that such antibodies generally display high affinity (nanomolar) for both components. (3) The serum antibody repertoire to the B HA displays a great degree of polarization with the top 5% highest frequency antibody clonotypes comprising majority (59-87%) of the repertoires both at days 28 and 180. (4) Pre-existing antibodies are much more likely to be cross-reactive, while new clonotypes emerging as a result of the vaccination tend to be more strain-specific. A number of serum antibodies were expressed recombinantly and analyzed for antigen specificity, HA inhibition and virus neutralization. The prevalence and relative abundance of neutralizing antibodies as a function of the magnitude and clonal diversity of the polyclonal response to the vaccine components will be presented.

AUTHOR INDEX

Abate, Adam R. ... 60
Abatemarco, Joe ... 111
Achenie, Luke ... 83
Acquaye, Edith ... 148
Acquaye-Seedah, Edith ... 133
Afshan, Gul ... 151
Almutairi, Zainab ... 130
Alper, Hal ... 111, 145
Alzbutas, Gediminas ... 152
Anderson, Mark ... 84
Arnold, Frances H. ... 131
Ataman, Meric ... 103
Badri, Abinaya ... 85
Baker, Jenny ... 94
Baldridge, Kevin ... 114
Barrick, Jeffrey E. ... 5
Barrios, Andres Fernando González ... 83
Beaujean, Pamphile ... 141
Bennett, R. Kyle ... 10
Bentley, William E. ... 76
Biggs, Bradley ... 12
Blanchard, Kristen ... 120
Blenner, Mark A. ... 105
Borges-Rivera, Diego ... 97
Bott, Michael ... 1
Boutz, Daniel R. ... 157
Brinkmann-Chen, Sabine ... 131
Broadbent, Andrew ... 150
Bryant-Friedrich, Amanda ... 114
Brynildsen, Mark P. ... 126
Bundy, Bradley C. ... 150
Byrom, Michelle ... 106
Cahn, Jackson K.B. ... 131
Cain, David ... 137
Calderon, Carlos ... 89
Calderon, Jose ... 89
Cardenas, Javier ... 11
Caschera, Filippo ... 8
Charab, Wissam ... 156
Chen, Anna H. ... 143
Chen, Linxiao ... 94
Chen, Yvonne ... 6
Chen, Zhilei ... 132
Cheong, Seokjung ... 128
Choi, Jin Wook ... 11
Cirino, Patrick C. ... 108, 138
Cochran, Jennifer R. ... 73
Contreras, Lydia ... 114
Contreras, Lydia M. ... 17, 118
Cramer, Shira ... 93, 95
Crook, Nathan ... 111
Cui, Lan-Yu ... 147
Da Silva, Nancy A. ... 11
Daringer, Nichole ... 74
Das, Aditi ... 12
Date, Amol ... 141
Daugherty, Patrick S. ... 101, 109
De Mey, Marjan ... 12
De Paz, Alexandra ... 135
De Soye, Ben ... 135
DeKosky, Brandon J. ... 156, 157
Del Cardayre, Steve ... 27
DeLisa, Matthew P. ... 94
Despaux, Leo ... 134
Detchanamurthy, Swaminathan ... 81
Digiovanni, John ... 93, 95
Dudek, Rachel M. ... 74
Dunn, Alexander ... 24
Ellington, Andrew D. ... 156
Ellington, Anrew ... 106
Elliott, Serra E. ... 101
Endicott, Christine ... 94
Fang, Kuili ... 112
Fast, Alan G. ... 10
Fengos, Geogios ... 103
Fernandez-Moya, Ruben ... 11
Fink, Gerald ... 9
Forkus, Brittany ... 22
Franco, Elisa ... 121
Frei, Christopher S. ... 108
Funa, Nobutaka ... 78
Futran, Alan ... 155
Gambill, Lauren ... 105
Gao, Yingning ... 100
Garcia, Juan ... 149
Gardiol, Daniel Hernandez ... 103
Ge, Xin ... 112
Geldart, Kathryn ... 22
Georgiou, George ... 93, 95, 156, 157
Gollihar, Jimmy ... 106
Gonzalez, Ramon ... 129
Gordonov, Tanya ... 76
Gostomski, Peter ... 81
Govindarajan, Sridhar ... 144
Grybauskas, Algirdas ... 152
Guan, Dongli ... 132
Guerrero, Jennifer L. ... 109
Hackel, Benjamin J. ... 71
Haitjema, Charles ... 97
Han, Songi ... 154
Harvill, Eric ... 133
Hassemer, Eryn ... 151
Hatridge, Taylor ... 118
Hatzimanikatis, Vassily ... 99, 103
Hauk, Pricila ... 141
Henske, John K. ... 97
Hodgman, C. Eric ... 84
Hoi, Kam Hon ... 156
Holec, Patrick ... 71
Hong, Seok Hoon ... 135

AUTHOR INDEX

Horton, Andrew P. 157
Huang, He 136
Huang, Jack Chung 94
Hunt, Jeremy 150
Ippolito, Gregory C. 156, 157
Jang, SungHo 140
Jayaraman, Guhan 85
Jensen, Michael K. 75
Jewett, Michael C. 77, 84, 135
Ji, Xiaojun 136
Jones, Bradley 94
Jung, Gyoo Yeol 7, 139, 140
Kaniusaite, Milda 152
Kanwar, Manu 141
Karlsson, Amy J. 125
Kaznessis, Yiannis N. 22
Keasling, Jay D. 75, 134
Kelleher, Neil 135
Khan, Tarik 80
Kiessling, Laura 3
Kim, Seohyoung 129
Kiparissides, Alexandros 99
Kojima, Takaaki 156
Kondo, Akihiko 4
Kruziki, Max A 71
Kwon, Yong-Chan 135
Kyriakou, Panagiota 22
Lagunavicius, Arunas 152
Lahiry, Ashwin 123
Lam, Felix H. 9
Lease, Richard A. 123
Leber, Christopher 11
Lee, Jiwon 156, 157
Leonard, Joshua N. 74
Lim, Chin Giaw 12
Lim, Hyun Gyu 139
Lim, Jae Hyung 139
Lin, Jyun-Liang 100
Link, A. James 155
Liu, Chang C. 28
Liu, Long 98
Lopez, Tyler 112
Lucks, Julius B. 92
Maass, Katie 104
Mahadevan, Radhakrishnan 16
Marcotte, Edward M. 157
Martin, Rey W 135
Mathur, Melina 119
Matsumura, Ichiro 120
Maynard, Jennifer 133, 148
Meijrink, Ben 153
Mienda, Bashir S. 90
Miskovic, Ljubisa 103
Moghaddam-Taaheri, Parisa 125
Moreno, Maria 149
Murray, Richard M. 121
Murrin, Ellen M. 157
Nam, Dong Hyun 112
Nguyen, Annalee 133
Nicholes, Nathan 141
Noireaux, Vincent 8
Ntai, Ioanna 135
O'Malley, Michelle A. 97, 109
Ostermeier, Marc 91, 141, 142
Padilla, Jose 149
Papin, James 133
Papoutsakis, Eleftherios T. 10
Parayil, Ajikumar 12
Park, Daechan 157
Payne, Gregory F. 76
Peebles, Christie A.M. 96
Pothukuchy, Arti 106
Powell, Warren 17
Putnam, David 94
Qian, Shuai 108
Quake, Stephen R. 157
Quan, David 76
Rajan, Saravanan 88
Rajkumar, Arun S. 75
Raman, Karthik 85
Redden, Heidi 145
Reddy, Sai T. 80
Reed, Jennifer 19
Reyes, Kris 17
Rivera-Silva, Gerardo 149
Robic, Srebrenka 120
Robinson, Jonathan L. 126
Rodin, Alexa 156
Rodriguez, Carlos 112
Romero, Philip A. 60
Roubos, Johannes A. 153
Runguphan, Weerawat 134
Saez, Rodrigo Torres 83
Saha, Achinto 93, 95
Salehi, Amin 150
Salgado, Adrian 122
Salis, Howard M. 25
San, Ka Yiu 154
Sandoval, Nicholas R. 10
Schinn, Matt 150
Schwarz, Kelly A. 74
Segura, Sonia Milena Aguilera 83
Seo, Sangwoo 140
Serber, Zach 26
Shabbir-Hussain, Murtaza 105
Shamsir, Mohd S. 90
Shelat, Nirav 142
Shi, Changhua 137
Shin, Jonghyeon 8
Shvartsman, Stanislav Y. 155

AUTHOR INDEX

Sidhu, Sachdev 88
Siegal-Gaskins, Dan 121
Skjødt, Mette 75
Smith, Mark T. 150
Smolke, Christina D. 119
Soffici, Alex R. 101
Solano, Oscar A. Alvarez 83
Solomon, Kevin V. 97
Stark, Jessica C. 84
Steinberg, Barrett 91
Stephanopoulos, Gregory 9
Stimple, Samuel D. 123
Stone, Everett 95
Stone, Everett M. 93
Styczynski, Mark P. 107
Sun, Jie 111
Tadi, Surendar 93, 95
Terrell, Jessica 76
Tervo, Christopher J. 19
Tessier, Peter M. 13
Thompson, Dawn A. 97
Tirrell, David 2
Tiziani, Stefano 93, 95
Tsao, Chen-Yu 76
Tullman-Ercek, Danielle 117
Van Deventer, James A. 88
Varadarajan, Navin 14
Vasquez, Jorge 17
Vasquez, Kevin 118
Vazquez-Anderson, Jorge 118
Velez, Vanessa Lucia Nuñez 83
Vlysidis, Michail 22
Voigt, Christopher A. 153
Vollmers, Christopher 157
Wagner, Ellen K. 133
Wang, Qinhong 110
Wang, Zhiqing 138
Watstein, Daniel 107
Wheeldon, Ian 100
Whitaker, W. Brian 10
Williams, Samuel 105
Wittrup, K. Dane 88, 104
Wolf, Roman 133
Wood, David W. 123, 137
Wu, Hsuan-Chen 76
Xing, Xin-Hui 147
Young, Eric M. 153
Zaman, Muhammad H. 21
Zargar, Amin 76
Zhang, Chong 147
Zhang, Guoqiang 110
Zhang, Jie 75
Zhang, Kechun 82
Zhang, Wenjun 18
Zhang, Xiaolin 19
Zhou, Hong 71
Zhou, Tiffany 121
Zhu, Jie 100
Zhu, Wen-Liang 147